Diversity of Service Academy Entrants and Graduates

T0195520

Sheila Nataraj Kirby, Harry J. Thie,
Scott Naftel, Marisa Adelson

Prepared for the Office of the Secretary of Defense
Approved for public release; distribution unlimited

NATIONAL DEFENSE RESEARCH INSTITUTE

The research described in this report was prepared for the Office of the Secretary of Defense (OSD). The research was conducted in the RAND National Defense Research Institute, a federally funded research and development center sponsored by OSD, the Joint Staff, the Unified Combatant Commands, the Department of the Navy, the Marine Corps, the defense agencies, and the defense Intelligence Community under Contract W74V8H-06-C-0002.

Library of Congress Cataloging-in-Publication Data is available for this publication.

ISBN 978-0-8330-4893-6

The RAND Corporation is a nonprofit research organization providing objective analysis and effective solutions that address the challenges facing the public and private sectors around the world. RAND's publications do not necessarily reflect the opinions of its research clients and sponsors.

RAND® is a registered trademark.

Published 2010 by the RAND Corporation
1776 Main Street, P.O. Box 2138, Santa Monica, CA 90407-2138
1200 South Hayes Street, Arlington, VA 22202-5050
4570 Fifth Avenue, Suite 600, Pittsburgh, PA 15213-2665
RAND URL: http://www.rand.org/
To order RAND documents or to obtain additional information, contact
Distribution Services: Telephone: (310) 451-7002;
Fax: (310) 451-6915; Email: order@rand.org

Preface

In the wake of concern about the diversity of candidates selected by the U.S. military service academies, House Report 110-279 requested that the Secretary of Defense conduct a comprehensive assessment of the recruiting efforts, admissions policies, graduation rates, and career success rates at the service academies. The military services provided data and accompanying materials pursuant to a request from the office of the Under Secretary of Defense for Personnel and Readiness. The RAND Corporation was asked to provide assistance in responding to the congressional mandate by summarizing trends in accession and various outcomes of interest for the entry cohorts as a whole and for subgroups of interest. The study was a short-term effort and limited in scope to descriptive analyses. This monograph should be of interest to personnel and military planners working to improve the diversity of the military services, particularly the officer force.

This research was sponsored by the Office of the Secretary of Defense and conducted within the Forces and Resources Policy Center of the RAND National Defense Research Institute, a federally funded research and development center sponsored by the Office of the Secretary of Defense, the Joint Staff, the Unified Combatant Commands, the Navy, the Marine Corps, the defense agencies, and the defense Intelligence Community. The principal investigators are Sheila Nataraj Kirby and Harry J. Thie. Comments are welcome and may be sent to Sheila_Kirby@rand.org or Harry_Thie@rand.org.

For more information on RAND's Forces and Resources Policy Center, contact the Director, James Hosek. He can be reached by email at James_Hosek@rand.org; by phone at 310-393-0411, extension 7183; or by mail at the RAND Corporation, 1776 Main Street, P.O. Box 2138, Santa Monica, California 90407-2138. More information about RAND is available at www.rand.org.

Contents

Figures

Tables

Summary

In the wake of concern about the diversity of cadets/midshipmen selected by the service academies, Congress requested that the Secretary of Defense conduct a comprehensive assessment of the recruiting efforts, admissions policies, graduation rates, and career success rates of entrants and graduates at the United States Military Academy (USMA), the United States Air Force Academy (USAFA), and the United States Naval Academy (USNA). RAND was asked to provide assistance in responding to the congressional mandate by summarizing trends in accession and various outcomes of interest for the entry cohorts as a whole and for subgroups of interest. This monograph documents RAND's analysis of the data provided by the military services pursuant to the congressional mandate.

The academies provide tuition-free, four-year undergraduate education and prepare entrants to be officers of the U.S. military services. Graduates are commissioned as officers for a minimum of five years. Each of the service academies admits between 1,100 and 1,350 entrants each year, and this has remained consistent over time and across the three academies.

Data and Methods

Data received from each academy included the following:

- number of students in the entry cohort by self-identified characteristics (race, ethnicity, and gender) for classes that entered the academy between 1992 and 2009
- number of students in each group who successfully completed the first year and transitioned to the second year for classes that entered the academy between 1992 and 2008 (2007, in the case of USAFA)
- number of students in each group who graduated from the service academy for classes that entered the academy between 1992 and 2005
- for the graduating classes of 1993–2003, data on rates of initial service obligation (ISO) completion and the percentage of graduates who remained in service as of June 2008.

All our data are as of June 30, 2008, so outcomes are defined as of that point in time. We also obtained data on the college graduation rates of other four-year institutions, against which we compared the graduation rates of the service academies.

Because our data are aggregated, we used simple tabulations for our analysis. We calculated outcomes in two ways: as a percentage of the entering class or graduating class, as appropriate, and as a percentage of those who successfully made it to the preceding outcome point (for example, graduation rate of those who successfully completed the first year) or conditional outcomes. To compare changes over time, we present three-year averages for the earliest and most recent cohorts for the outcomes of interest; averaging also alleviates problems related to some groups' small sample sizes.

We tested for statistically significant differences in outcomes across all years by race/ethnicity, gender, and cohort year. Some estimated differences were small (although significant); as a result, we focus on the more substantive findings here. In addition, we should note that these statistically significant differences in outcomes are across all cohorts. The data on recent cohorts show improvement in first-year completion and graduation rates; thus, significant differences may not exist in the future.

An important limitation of our data on continuation rates beyond ISO is that, for a given year beyond graduation, we have data on the continuation rate of only *one* graduating class. Thus, it is difficult to generalize to other classes with respect to the percentage likely to stay in service beyond ISO.

United States Military Academy

Table S.1 presents summary data for the three earliest and the three most recent USMA cohorts for the outcomes of interest.

Demographic Profile of Entrants

The percentage of women in the three most recent entering classes (2007–2009) was 16 percent, higher than the 12 percent in the earliest entering classes for which we have data (1992–1994). The percentage of nonwhites has increased over time, from 16 percent in the 1992–1994 entering classes to 23 percent in the 2007–2009 entering classes. Of the 2007–2009 entering classes, 6 percent were black or African-American, 9 percent were Hispanic or Latino, 7 percent were Asian, and 1 percent were American Indian or Alaska Native.

First-Year Completion and Graduation

The percentage successfully completing the first year increased from 84 percent in the earliest classes to 91 percent in the most recent classes. Women, in particular, increased their first-year completion rate by 10 percentage points. Every racial/ethnic group increased its first-year completion rate by 3–10 percentage points.

Graduation rates remained relatively constant over time (76–77 percent). Again, women increased their graduation rates by 5 percentage points (from 69 percent to 74 percent). With one exception (Asians), all racial/ethnic groups either maintained or slightly increased their graduation rates over time. The graduation rate conditional on successfully completing the first year (not shown in the table) declined from 92 percent among the earliest cohorts (1992–1994) to 87 percent among the 2003–2005 entering cohorts.

Table S.1
Profile of Entrants and Selected Outcomes, Three Earliest and Three Most Recent Cohorts, USMA

Characteristic	Three Earliest Cohorts (%)	Three Most Recent Cohorts (%)
Entrants[a]		
Women	12	16
Nonwhite	16	23
First-year completion rate[b]		
Average	84	91
Gender		
Men	85	91
Women	80	90
Race/ethnicity		
Asian	85	95
Black or African-American	87	90
White	84	91
Hispanic or Latino	85	92
Graduation rate[c]		
Average	77	76
Gender		
Men	78	77
Women	69	74
Race/ethnicity		
Asian	80	76
Black or African-American	74	74
White	77	77
Hispanic or Latino	74	76

Table S.1—Continued

Characteristic	Three Earliest Cohorts (%)	Three Most Recent Cohorts (%)
ISO completion rates of graduates[d]		
Average	82	91
Gender		
Men	82	92
Women	79	91
Race/ethnicity		
Asian	82	94
Black or African-American	80	88
White	82	92
Hispanic or Latino	82	92

[a] Comparing 1992–1994 to 2007–2009 entering classes.
[b] Comparing 1992–1994 to 2006–2008 entering classes.
[c] Comparing 1992–1994 to 2003–2005 entering classes.
[d] Comparing 1993–1995 to 2001–2003 graduating classes.

ISO Completion

Graduates' rate of ISO completion increased by 9 percentage points over time (from 82 percent for the 1993–1995 graduating classes to 91 percent for the 2001–2003 graduating classes). About 91 percent of women graduates completed their ISO in the most recent cohorts, compared with 79 percent in the earliest cohorts. The increase in ISO completion over the same period ranged from 8 percentage points for blacks to 12 percentage points for Asians.

Continuation Rates

Of the graduating class of 2003, 73 percent remained in service as of June 2008, when they had just completed their ISO.[1] For the graduat-

---

[1]   As mentioned, our data on continuation rates are limited because each data point represents the experience of only one graduating class. Thus, we do not show them in the tables.

ing class of 2002, there was a sharp drop-off in the continuation rates of graduates: Only 52 percent remained in service one year beyond their ISO. If the experiences of the graduating classes are similar over time, then we would expect one-third of graduates to remain in the military for seven to ten years beyond their ISO (12–15 years beyond graduation).

Women had much lower continuation rates than men, and this was true of every graduating class. The continuation rates of nonwhite graduates tended to be more variable, but, in general, they—especially Hispanics—appeared to continue at higher rates than whites six to nine years beyond their ISO.

United States Air Force Academy

Demographic Profile of Entrants

As shown in Table S.2, the percentage of women in the USAFA increased from 15 percent in the earliest classes for which we have data to 21 percent in the 2007–2009 entering classes. The percentage of non-whites increased from 18 percent to 22 percent over the same period. In the most recent classes, 6 percent of academy entrants were black or African-American, 8 percent were Hispanic or Latino, 7 percent were Asian, and 1 percent were American Indian or Alaska Native.

First-Year Completion and Graduation Rates

The first-year completion rate increased from 81 percent in the three earliest cohorts to 85 percent in the three most recent cohorts. Women increased their first-year completion rate only slightly (by 1 percentage point). All racial/ethnic groups increased their first-year completion rate—Hispanics by 2 percentage points, blacks by 8 percentage points, and Asians by 6 percentage points.

The graduation rate also increased over time, from 72 percent in the three earliest cohorts to 76 percent in the most recent cohorts. Women posted a gain of 3 percentage points, while blacks and Hispanics increased their graduation rate by 2 percentage points. Asians and whites experienced larger increases of 6–7 percentage points.

Table S.2
Profile of Entrants and Selected Outcomes, Three Earliest and Three Most Recent Cohorts, USAFA

Characteristic	Three Earliest Cohorts (%)	Three Most Recent Cohorts (%)
Entrants[a]		
Women	15	21
Nonwhite	18	22
First-year completion rate[b]		
Average	81	85
Gender		
Men	81	86
Women	81	82
Race/ethnicity		
Asian	85	91
Black or African-American	81	89
White	80	84
Hispanic or Latino	84	86
Graduation rate[c]		
Average	72	76
Gender		
Men	72	76
Women	72	75
Race/ethnicity		
Asian	69	76
Black or African-American	70	72
White	71	77
Hispanic or Latino	73	75

Table S.2—Continued

Characteristic	Three Earliest Cohorts (%)	Three Most Recent Cohorts (%)
ISO completion rates of graduates[d]		
Average	90	82
Gender		
Men	91	84
Women	86	69
Race/ethnicity		
Asian	93	80
Black or African-American	90	76
White	90	82
Hispanic or Latino	88	84

[a] Comparing 1992–1994 to 2007–2009 entering classes.
[b] Comparing 1992–1994 to 2005–2007 entering classes.
[c] Comparing 1992–1994 to 2003–2005 entering classes.
[d] Comparing 1993–1995 to 2001–2003 graduating classes.

As was true for the USMA cohorts, the conditional graduation rate declined slightly, by 2 percentage points, between the two periods (not shown); we see declines in conditional graduation rates among women, blacks, and Hispanics.

ISO Completion Rates

Among graduates, the ISO completion rate declined from 90 percent for the earliest cohorts to 82 percent for the more recent cohorts. The ISO completion rate declined markedly for women (from 86 percent to 69 percent) and for blacks (from 90 percent to 76 percent) and Asians (from 93 percent to 80 percent). Exogenous factors—such as reductions in force or the civilian economy—are likely to affect retention, and this may help explain the decline. An analysis of the effects of such factors was beyond the scope of our study.

Continuation Rates

Of the graduating class of 2003, 82 percent remained in service as of June 2008, when they had just completed their ISO. Over time, we found that about half of the graduates stayed 6–10 years beyond their ISO. This continuation rate was higher than that of the USMA cohorts, largely because some graduates incur additional service obligations as pilots. Women had much lower continuation rates than men, and this was true of every graduating class. Blacks in most years continued in service at lower rates than other groups.

United States Naval Academy

Demographic Profile of Entrants

As with the USAFA, the percentage of women in the USNA increased from 15 percent in the earliest classes to 21 percent in the 2007–2009 entering classes (Table S.3). The percentage of nonwhites increased from 18 percent to 22 percent over the same period. In the most recent classes, 5 percent of academy entrants were black or African-American, 12 percent were Hispanic or Latino, 4 percent were Asian, and less than 1 percent were American Indian or Alaska Native. The percentage of nonwhites in the 2009 entering class had risen to 28 percent.

First-Year Completion and Graduation Rates

The first-year completion rate increased by 6 percentage points (from 88 percent in the three earliest cohorts to 94 percent in the three most recent cohorts). Women increased their first-year completion rate by 8 percentage points to 91 percent in the most recent cohorts. All racial/ethnic groups increased their first-year completion rate, and there was little difference in completion rates across the various groups.

The graduation rate also increased over time, from 78 percent in the three earliest cohorts to 85 percent in the most recent cohorts, the highest among the three academies. Women increased their graduation rates substantially, from 70 percent to 85 percent over this period, bringing it on a par with that of men. Hispanics also experienced a

Table S.3
Profile of Entrants and Selected Outcomes, Three Earliest and Three Most Recent Cohorts, USNA

Characteristic	Three Earliest Cohorts (%)	Three Most Recent Cohorts (%)
Entrants[a]		
Women	15	21
Nonwhite	18	22
First-year completion rate[b]		
Average	88	94
Gender		
Men	89	95
Women	83	91
Race/ethnicity		
Asian	90	93
Black or African-American	87	93
White	88	94
Hispanic or Latino	86	94
Graduation rate[c]		
Average	78	85
Gender		
Men	79	85
Women	70	85
Race/ethnicity		
Asian	83	91
Black or African-American	67	75
White	79	86
Hispanic or Latino	68	81

Table S.3—Continued

Characteristic	Three Earliest Cohorts (%)	Three Most Recent Cohorts (%)
ISO completion rates of graduates[d]		
Average	95	89
Gender		
Men	95	91
Women	89	75
Race/ethnicity		
Asian	96	83
Black or African-American	94	83
White	95	90
Hispanic or Latino	94	86

[a] Comparing 1992–1994 to 2007–2009 entering classes.
[b] Comparing 1992–1994 to 2006–2008 entering classes.
[c] Comparing 1992–1994 to 2003–2005 entering classes.
[d] Comparing 1993–1995 to 2001–2003 graduating classes.

marked increase in graduation rates—from 68 percent to 81 percent—while blacks increased their graduation rates by 8 percentage points to 75 percent. Asians and whites also posted increases of 7–8 percentage points in graduation rates.

Women in the earlier earlier cohorts had higher conditional graduation rates than men; in the most recent cohorts, the rate for men increased slightly by 2 percentage points to 91 percent, bringing them on a par with women. The conditional graduation rates of blacks and Hispanics improved by 8–10 percentage points to 85 percent and 89 percent, respectively.

ISO Completion Rates

Among graduates, the ISO completion rate declined, from 95 percent for the earliest cohorts to 89 percent for the more recent cohorts. As with the Air Force, the ISO completion rate declined markedly for women (from 89 percent to 75 percent). We see large declines of 8–13 percentage points in ISO completion rates among all nonwhite groups. As mentioned earlier, these rates are likely to be affected by service policies, such as reductions in force, or by competition from the civilian economy.

Continuation Rates

Of the graduating class of 2003, 86 percent remained in service as of June 2008, when they had just completed their ISO. For the graduating class of 2002, 82 percent remained in service one year beyond their ISO. Over time, 40–50 percent of graduates stayed seven to ten years beyond their ISO.

Women had lower continuation rates than men, and this was true of every graduating class. Hispanics generally continued in service at rates similar to those of whites, although blacks in most years continued at lower rates than other groups.

USNA Graduates Who Join the Marine Corps

Demographic Profile. On average, about 17 percent of USNA graduates between 1996 and 2003 joined the Marine Corps after graduation (the number ranged from 148 to 165). Of the 2003 USNA graduates who joined the Marine Corps, 11 percent were nonwhite and 13 percent were women (somewhat lower than in the preceding two classes).

ISO Completion. USNA graduates who joined the Marine Corps had a higher rate of ISO completion than those who joined the Navy, regardless of graduating class. In the three most recent cohorts (2001–2003), the average rate of ISO completion was 97 percent for Marine Corps officers, compared with 87 percent for Navy officers.

Continuation Rates. With the exception of the 2003 graduating class, Navy officers tended to have slightly higher continuation rates than Marine Corps officers in most years.

Statistically Significant Differences in Selected Outcomes Across All Cohorts

Table S.4 summarizes statistically significant differences among groups for the selected outcomes across the three service academies. These data are across all years; the table does not simply compare the earliest and most recent cohorts, so some of these differences may not hold in future cohorts if outcomes for selected groups continue to improve and differences between groups become smaller. For the sake of brevity, we refer to statistically significant differences as *significant differences*.

- In seven of the nine outcomes considered here, across all cohorts, women had outcomes that were statistically different and lower than those of men. However, in all three academies, as noted earlier, women improved both their first-year completion and graduation rates to the point that the differences between men and women were substantially reduced or eliminated altogether. However, in recent cohorts, USAFA and USNA women graduates have had markedly lower ISO completion rates than men.
- Asians generally had outcomes similar to or better than those of whites. However, in recent cohorts, Asians graduating from USAFA and USNA have had lower rates of ISO completion than whites.
- Blacks had significantly lower graduation rates than whites in USMA and USNA, and, despite recent increases, this continues to be the case. In recent cohorts, the differences in graduation rates are 3 percentage points in USMA, 5 percentage points in USAFA, and 11 percentage points in USNA. Blacks also tended to have lower ISO completion rates, and the differences are even more marked in recent cohorts.
- Hispanics have had significantly lower graduation rates than whites across all three academies. However, in recent cohorts, they have closed the gap to 1 percentage point in USMA, 2 percentage points in USAFA, and 5 percentage points in USNA.

Thus, while looking across cohorts is useful, it is important to pay attention to the experience of the more recent cohorts to determine priorities for investing resources to improve outcomes for diverse groups.

Table S.4
Statistically Significant Estimated Differences in Selected Outcomes Across All Years, by Gender and Race/Ethnicity

Characteristic	First-Year Completion	Graduation from the Academy	Completion of ISO
Women, compared to men			
USMA	Lower	Lower	—
USAFA	*Lower*	—	Lower
USNA	Lower	Lower	Lower
Compared to whites			
USMA			
Asian	Higher	Higher	Higher
Black or African-American	—	Lower	*Lower*
Hispanic or Latino	—	Lower	—
USAFA			
Asian	Higher	—	—
Black or African-American	—	—	Lower
Hispanic or Latino	—	Lower	—
USNA			
Asian	—	—	*Lower*
Black or African-American	—	Lower	Lower
Hispanic or Latino	*Lower*	Lower	*Lower*

NOTE: All cell entries represent statistically significant differences relative to the reference group. Entries in italics are those for which the odds ratio was modest or the confidence interval was wide (the upper or lower limit was close to 1), indicating a weak effect.
— indicates no significant difference between groups.

Comparing Academy Graduation Rates to Graduation Rates of "Very Selective" Four-Year Institutions

We compared the graduation rates of the service academies to those of "very selective" civilian four-year institutions, using data published by the National Center for Education Statistics. There are two points to note about the comparisons: First, the civilian institution data shown are for the freshman entering class of 1998, while the service academy data are aggregated across the 1992–2005 entering classes; second, entrants typically graduate from the service academies in four years, so we are comparing four-year graduation rates to six-year graduation rates for civilian institutions. In the 1998 freshman class, the percentage of women enrolled in very selective civilian four-year institutions in the study (n = 117) was 57 percent, much higher than the 16–21 percent enrolled in the most recent academy entry cohorts. The percentage of nonwhites enrolled in the freshman class, however, was similar between the civilian and military institutions—23 percent in the civilian institutions compared with 22–23 percent in the most recent academy entry cohorts.[2]

Figure S.1 shows the graduation rates by race/ethnicity and gender for very selective four-year institutions and for the service academies. Academy graduation rates are higher than those in comparable civilian four-year institutions on average and across all racial/ethnic groups. For example, 72 percent of blacks graduated from the service academies, on average, compared with 60 percent who attended four-year civilian institutions—a substantial 12-percentage-point gap in graduation rates. We noted earlier that graduation rates for the most recent cohorts entering the academies (2003–2005) have increased; if this improvement is sustained, the gap in graduation rates may be even larger.

There is a 3-percentage-point difference in the graduation rates of women (74 percent versus 77 percent in the academies and civilian institutions, respectively) and a 5-percentage-point difference in the graduation rates of men (78 percent versus 73 percent). However, while

---

[2] Using data only for 1998 academy entry cohort, we found that the percentage of women was slightly smaller (15–16 percent), as was the percentage of nonwhites (18–19 percent), than in the most recent cohorts.

Figure S.1
Six-Year Graduation Rates in Very Selective Four-Year Institutions, 2004,
and the Service Academies, 2003–2005, by Race/Ethnicity and Gender

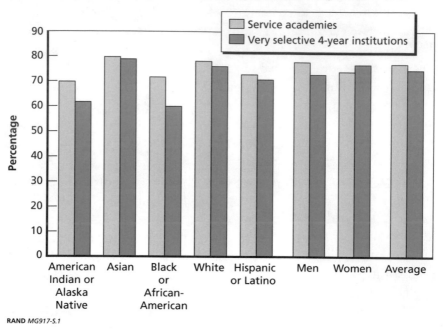

RAND *MG917-S.1*

women have lower graduation rates than men in the academies, the opposite is true in the civilian institutions. We should note that the difference in graduation rates between men and women in the most recent graduating cohort declined to 1 percentage point.

Service Action Plans

Each of the military departments has action plans detailing ways to improve diversity and representations.

USMA. The primary minority recruitment tool is Project Outreach, which seeks to identify and nurture talented minority candidates through the admissions process with the ultimate goal of matriculating them to West Point. Other programs include weekend visits of prospective recruits to USMA and the United States Military Academy

Prep School; visits with the Congressional Black and Hispanic Caucuses to set up "academy days" and place cadets as interns in local and Washington, D.C., offices; minority cadets' participation in hometown visits and academy days; and the Cadet Calling Program, through which current cadets interact with candidates via phone. Several other new initiatives are under way, including examining best practices at other tier-1 institutions.

USAFA. To identify candidates, USAFA will advertise in prominent minority and urban media outlets and increase emphasis on coordinators who help identify, mentor, and evaluate diverse candidates. Among other initiatives, it proposes to offer one-week summer seminars between junior and senior years, expand the diversity visitation program to bring applicants to USAFA for a visit, and provide support to cadets of diverse backgrounds to help ensure their success.

USNA. A new diversity office, led by a senior naval officer, was created and staffed to be the single coordinating entity for all diversity efforts.

Recommendations for the U.S. Department of Defense to Support Service Efforts to Improve Diversity

The action plans adopted by the service academies encompass several specific strategies. At a higher level, the U.S. Department of Defense (DoD) and the services need to take steps both to support these plans and to ensure that the plans are linked to the larger DoD vision and goals. Specifically, the Office of the Secretary of Defense should pursue the following initiatives:

- Review and communicate DoD's definition of *diversity*.
- Determine what needs to be measured according to the leadership's vision and mission for diversity and the best metrics for that purpose.
- Review goals for diversity and ensure that they are aligned with DoD's overall mission.

- Emphasize that diversity management is a priority for the entire organization and has the backing of the highest level of DoD leadership, not merely the personnel community.
- Focus efforts not simply on accessing a more diverse group of officers but on increasing career retention of these officers.

Acknowledgments

We thank Curtis Gilroy, Director of Accession Policy, and Lt Col Rose Jourdan, our sponsors, for their interest in and support of this project. We also thank the service academies for providing the data for this study and for useful comments on an earlier draft of this monograph. Additional comments provided by the Office of General Counsel (Personnel and Health Policy) were helpful in improving the clarity of the final product.

We are grateful to our RAND reviewers—Heather Krull and Nelson Lim—for their thoughtful and constructive reviews and helpful suggestions for improving the readability of the monograph. We are particularly grateful to Heather, who provided very useful comments on a revised version of the draft, despite tight time constraints. The final document is considerably improved as a result of their suggestions and recommendations. We thank Lauren Skrabala, our editor, for her careful and skillful editing, Mary Wrazen and Carol Earnest for their excellent work on the figures, and Steve Oshiro, our production editor, for his efficient oversight of the publication process.

Abbreviations

CLF	Civilian Labor Force
DMDC	Defense Manpower Data Center
DoD	U.S. Department of Defense
EEOC	U.S. Equal Employment Opportunity Commission
FY	fiscal year
GAO	U.S. Government Accountability Office
HENAAC	Hispanic Engineers National Achievement Awards Corporation
IPEDS	Integrated Postsecondary Education Data System
ISO	initial service obligation
NAACP	National Association for the Advancement of Colored People
NASA	National Aeronautics and Space Administration
NAVCO	Navy Office of Community Outreach
NCES	National Center for Education Statistics
OMB	Office of Management and Budget
OPM	Office of Personnel Management

ROTC	Reserve Officers' Training Corps
USAFA	United States Air Force Academy
USMA	United States Military Academy
USNA	United States Naval Academy

CHAPTER ONE

Introduction

The U.S. Congress has expressed concern about the diversity of officers selected by the U.S. military service academies. House Report 110-279 requested that the Secretary of Defense conduct a comprehensive assessment of the recruiting efforts, admissions policies, graduation rates, and career success rates with respect to the diversity of entrants and graduates at the service academies: the United States Military Academy (USMA), the United States Air Force Academy (USAFA), and the United States Naval Academy (USNA). This monograph examines data on the academies' classes over the past ten years to evaluate past trends and current standings. The Secretary of Defense was also asked to provide an action plan detailing ways to improve diversity and representation among the nation's service academies, as appropriate.

There is concern that officers from different commissioning sources have different propensities to stay in the military and to make the military a career. This has focused attention on the criteria that these different commissioning sources use to select individuals at entry and the relationship between these criteria and future career retention and performance.[1] One concern is that these criteria may exclude a portion of the youth population that has the ability to succeed academically and could possess other desirable attributes, such as commitment to the military.

RAND was asked to provide assistance to the office of the Under Secretary of Defense for Personnel and Readiness, both in responding

---

[1] Previous RAND work examined differences in career progression and continuation rates among junior officers by race/ethnicity and gender. See Hosek et al. (2001).

to the congressional mandate and in conducting a larger study of the selection criteria used for admission into the service academies and the award of Reserve Officers' Training Corps (ROTC) scholarships and their relationship to career retention and performance.

This monograph documents RAND's analysis of the data provided by the military services pursuant to the congressional mandate. It should be noted that this monograph is limited in scope and does not fully address all the issues raised in the congressional request. While we examined the gender and race/ethnicity of academy entrants and graduates, we did not analyze their national origin or religion. The general consensus was that these data were likely to be either missing for a large number of entrants or not very reliable. In addition, we did not examine the current ranks of graduates remaining in service. Because most of the graduating cohorts have not had sufficient time to receive meritorious promotions, it is unlikely that we would find meaningful differences among graduates with respect to rank.

We also include some suggested steps that the U.S. Department of Defense (DoD) could take to support the services' action plans to improve diversity in the academies. These suggestions are based on recently published RAND work and other literature. Data on diversity in the service academies are placed in the context of the diversity of the military officer force, of civilian college graduates aged 21–35 years, and of civilian college graduates aged 21–49 years in the civilian workforce.

Background on the Service Academies

The three service academies—USMA, USAFA, and USNA—are overseen by the three military departments of DoD (the Army, Air Force, and Navy, respectively). These institutions provide tuition-free, four-year undergraduate education and prepare entrants to be officers of the U.S. military services. Upon graduation, individuals are commissioned as officers in the military for a minimum of five years.

To be considered for an appointment to a service academy, an applicant must meet the eligibility requirements established by law and be nominated by an authorized person, such as a member of the U.S. Senate or House of Representatives. Allocations for nominations of prospective appointees by members of Congress are governed by statute (10 U.S.C. 4342, USMA; 10 U.S.C. 9342, USAFA; and 10 U.S.C. 6954, USNA). The numbers of positions subject to congressional nomination are as follows:

- ten from each state, five of whom are nominated by each senator from that state
- five from each congressional district, nominated by the representative from the district
- five from the District of Columbia, nominated by the delegate from the District of Columbia
- two to five from the U.S. Virgin Islands, Puerto Rico, Guam, and American Samoa.

An applicant for a nomination must meet the following eligibility requirements as of July 1 of the year of admission to a service academy:

- U.S. citizenship
- at least 17 years of age and not yet 23 years old on July 1 of the year the applicant would enter an academy
- unmarried
- not pregnant and without legal obligation to support children or other dependents (USMA, undated; USAFA, undated; USNA, undated).

In addition to these requirements, each academy has established academic, physical, and leadership requirements for admission. As noted earlier, acceptance of a service academy appointment typically requires at least a nine-year commitment, including four years at the academy and five years of an active-duty initial service obligation (ISO). Cadets/midshipmen do not incur a service obligation until the beginning of their junior year.

Each of the service academies admits between approximately 1,100 and 1,350 entrants each year, and this has remained consistent over time and across the three academies.

Data

The three military service academies (USMA, USNA, and USAFA) were provided a template requesting data on all academy entrants who graduated between 1996 and 2009 or who would graduate between 2010 and 2013. The data included the 1992–2009 entering classes or entry cohorts. The academies were asked to provide information on these entry cohorts at various points—first-year continuation rates, graduation rates, rates of ISO completion, and the percentage of graduates remaining in service as of June 30, 2008. These latter two outcomes were included only for those who graduated earlier than 2004.

We received the following data from the academies:

- number of students in the entry cohort by self-identified characteristics (race, ethnicity, and gender)[2]
- number of students in each group who successfully completed the first year and transitioned to the second year
- number of students in each group who graduated from the service academy.

Congress was also concerned with the return on academy education in terms of ISO completion and retention beyond the term of initial commitment. As noted earlier, the minimum active-duty service commitment for academy graduates is five years. Thus, for those graduating between 1993 and 2003, the service academies were also asked to provide data on

---

[2]   As noted earlier, we did not request data on the national origin or religion of entrants because of the general consensus that these data were likely to be missing for a large number of entrants or not very reliable. In addition, we did not ask for the current ranks of graduates. As of this writing, there has not been sufficient time for these officers to receive meritorious promotions, making it unlikely that we would find meaningful differences in rank.

- number of graduates (disaggregated by race, ethnicity, and gender) who completed their ISO as of June 30, 2008
- number of graduates who remained in service as of June 30, 2008.

To provide context for these data, we also obtained historical data from DoD on the diversity of all officer accessions and all officers by service over an 11-year period (1997–2007). Each year since 1975, DoD has reported to Congress on the population representation of the military services. This monograph provides information on the demographic characteristics of active-duty applicants, active and reserve enlisted accessions and members, and active and reserve officer accessions and officer corps. These data were provided by the Defense Manpower Data Center (DMDC). Historical reports and data are available at Deputy Under Secretary of Defense for Military Personnel Policy (undated). These reports also provide demographic data on all civilian college graduates (aged 21–35 years) and all civilian college graduates aged 21–49 years who are in the workforce. These are the comparison groups used for officer accessions and for the officer corps. In addition, we present data on the graduation rates of different groups of students enrolled in four-year institutions as context for the graduation rates achieved by the service academies.

One point to note is that, in fiscal year (FY) 2003, there was a change in the way race and ethnicity data were calculated and reported, pursuant to the new guidelines published by the Office of Management and Budget (OMB) (1997). Prior to January 2003, race categories were black, white, and other. The "other" category included Asian/Pacific Islander and American Indian. Those who identified themselves as "Hispanic" were defined as such, without regard to race, and included as a discrete category within all race/ethnicity tables. From 2003 on, DoD agencies were required to offer respondents the following five race categories:

- American Indian or Alaska Native
- Asian
- black or African-American

- Native Hawaiian or other Pacific Islander
- white.

Respondents could also check more than one category, and this was reported as "two or more races."

A separate question asked respondents their Hispanic identity:

- Hispanic
- not Hispanic.

In addition, FY 2003 was the first year for which we had data based on the 2000 decennial census to estimate characteristics of the civilian population. That census provided a more accurate estimate of the proportion of Hispanics and other minorities in the civilian population than the 1990 census. As a result of this change, the estimated proportion of Hispanic youth in the civilian population was approximately 2 percentage points higher in the FY 2003 report than the estimates found in earlier editions. Therefore, representation of Hispanics among military accessions and members will appear reduced due to the increase in the estimated ethnic composition of the civilian comparison population.

Despite this change, the academies were able to provide consistent race/ethnicity data across the years.

Methods

It is important to keep in mind that this study was a short-term effort and aimed to provide a descriptive analysis of trends in outcomes of academy entrants and graduates by gender and race/ethnicity. Clearly, a number of factors affect success at the academies, ISO completion, and continuation in service beyond the ISO point. Among others, these factors include mentoring and assistance provided to cadets/midshipmen at the academies, warfare specialty that might be linked to promotion opportunities, frequency of deployment, reductions in end strength implemented by the services, competition from the civilian economy, performance of the individual, and perceived discrimination. Analyzing the effects of these and other factors in a full-blown modeling effort was beyond the scope of the study, but these factors should be considered when examining the trends reported here.

Because our data are aggregated, we used simple tabulations to analyze them. In Chapter Two, we simply present an average snapshot of the racial/ethnic and gender profile of all officer accessions and the officer corps by service. Because of the change in the reporting of race/ethnicity data in 2003, we averaged data for entering cohorts before 2003 and for those entering from 2003 onwards. We show similar averages for all civilian college graduates aged 21–35 years and for all employed civilian college graduates aged 21–34 years. These are the groups that DoD has traditionally used to compare the diversity of its officer accessions and its officer corps.

In subsequent chapters that examine data on the service academies, we analyze the data in some detail. First, we report trends over time in the gender and racial/ethnic profile of entering cohorts. To compare changes over time, we averaged data on the three earliest and three most recent cohorts for which we had data. Second, we calculated averages for the outcomes of interest for the three earliest and three most recent cohorts by race/ethnicity and gender to track changes over time, although we show year-by-year trends as well. Averaging three years of data helps eliminate the large fluctuations that can result when calculating percentages for groups with small sample sizes, which is important to keep in mind when examining the year-by-year data in later chapters. For example, because the American Indian or Alaska Native group is very small, we do not show disaggregated data by year. But other groups also have small sample sizes, particularly in some years, so readers should be aware of that when looking at yearly trends. In the text, we focus on the three-year averages of the earliest and most recent cohorts when discussing trends over time.

We calculated first-year completion and graduation rates for entering classes; we also calculated graduation rates conditional on successfully completing the first year. To examine ISO completion and continuation in service beyond the ISO point, we used graduating classes and calculated percentages of graduates who completed

ISO and remained in service at different points in time. Because some graduates do not complete their ISO for a variety of reasons (including, for example, reductions in force implemented by the services), we also calculated continuation rates for graduates who had completed their ISO. These rates are reported in Appendix A.

One caveat must be mentioned. A real limitation of our data is that all we have is a snapshot of how many officers graduated in a particular year and were still in service as of June 30, 2008. Thus, for a given year beyond graduation, we have data on the continuation rate of *one* graduating class. So, for example, for the cohort that graduated in 2002, we can look at continuation rates six years beyond graduation; for the 2001 graduating class, we can look at continuation rates seven years beyond graduation. If we make the assumption—which may be questionable—that each cohort behaves exactly like another, then we can combine these data points to talk about "survival" probabilities over time. Because we observe only one cohort for each data point and the data were aggregated, we are unable to use statistical techniques, such as the Kaplan-Meier survival curves, to estimate these probabilities more precisely.

In all cases, we are interested in whether and how outcomes differ among different racial/ethnic groups or between men and women. Pairwise or multiple comparison tests would allow us to examine whether the differences we observe in different outcomes among different groups or different cohorts are statistically significant (at least for those groups for which the sample sizes are large enough). However, conducting multiple significance tests increases the chances of obtaining spurious significant results between some pairs of groups or cohorts. Instead, we used exploratory logistic regression models (for binned or aggregated data) to test in a multivariate framework whether there were any statistically significant differences among groups (or cohorts). We examined the width of the 95-percent confidence interval around the estimated odds ratios to see whether the relationship, despite being significant, was weak (for example, if the upper or lower limit of the confidence interval were close to 1). Because the data were limited, we could not carry out more sophisticated modeling and do not report the actual coefficients or full models here. The purpose was simply to understand whether the differences among groups were significant.

Organization of This Monograph

To set the context for the analyses reported in the next several chapters, Chapter Two provides an overview of the diversity of all officer accessions and all civilian college graduates aged 21–35 years. We also compare the diversity of the total officer corps with that of all civilian college graduates aged 21–49 years who are in the civilian workforce. A third section of Chapter Two presents data on the six-year graduation rates of students enrolled in four-year institutions based on a report by the National Center for Education Statistics (NCES).

Against this background, Chapters Three through Five provide a look at trends in the representativeness of entry cohorts into the service academies and at various outcomes, including first-year completion, graduation from the academy, completion of ISO, and continuation in the service beyond the ISO point. Chapter Six describes the service academies' own action plans for improving the diversity of admissions, and Chapter Seven contains selected findings and recommendations from the literature on improving diversity in organizations. Chapter Eight presents conclusions and outlines some steps that DoD might consider to support the services in their efforts to improve diversity, both in the academies and in the officer corps.

Appendix A presents an alternative way of looking at continuation rates: It displays the continuation rates by gender and race/ethnicity using graduates who had completed their ISO as the denominator. Because ISO completion rates tend to be quite high, the two sets of rates are not very different but offer another perspective on continuation. Appendix B provides a brief overview of active-duty service obligations for selected education and training opportunities.

CHAPTER TWO

Diversity of the Officer Corps and All Officer Accessions: Trends Over Time

This chapter presents data on the number and demographic makeup of the officer corps and all officer accessions by service from 1997 to 2007, as well the demographic profile of their traditional comparison groups—all employed civilian college graduates aged 21–49 years and all college graduates aged 21–35 years. A third section of this chapter presents data on civilian college graduation rates to provide context for the academy graduation rates discussed later.

All Officers

The total number of officers was a little over 212,000 in 1997, and this declined by about 4 percent to approximately 204,000 by 2007. Figure 2.1 shows the total number of officers by service over the period 1997–2007. Over this period, both the Army and the Marine Corps increased the size of their officer corps. The Army increased from approximately 68,000 officers in 1997 to 71,000 in 2007, and the Marine Corps increased from 16,000 officers to a little less than 18,000. The total number of Navy officers declined from a little over 54,000 to approximately 50,000 during this period, and the Air Force posted the largest decline—from 74,000 to 66,000 officers.

As a percentage of the total, the Army accounts for about a third of the officer corps (35 percent in 2007), the Navy for about one-quarter (24 percent in 2007), and the Air Force for about one-third (32 percent in 2007). The Marine Corps is the smallest of the services, accounting for about 8 percent of the officer corps (9 percent in 2007).

Figure 2.1
Total Number of Officers, by Service, 1997–2007

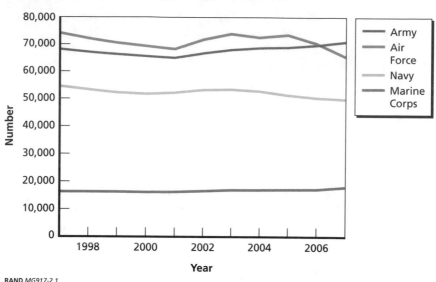

RAND *MG917-2.1*

Gender Representation

Figure 2.2 shows the percentage of women among officers by service and averaged across DoD, as well as among employed civilian college graduates aged 21–49 years, averaged over the 11 years. On average, during this period, women accounted for 15 percent of all officers across DoD: 15 percent of Army and Navy officers, 5 percent of Marine Corps officers, and 17 percent of Air Force officers.

The percentage of women has risen slightly in every service from 1997 to 2007: The percentage of women in the Army officer corps increased from 14 percent to 17 percent; in the Navy, from 14 percent to 15 percent; in the Marine Corps, from 4 percent to 6 percent; and in the Air Force from 16 percent to 18 percent.

Women accounted for about half of all employed civilian college graduates aged 21–35 years in 2007, an increase from 48 percent in 1997. The fact that the services have far smaller percentages of women than the civilian college graduate workforce is not unexpected—women are typically underrepresented in nontraditional

Figure 2.2
Percentage of Women in the Officer Corps, by Service, and Among
Employed Civilian College Graduates, 21–35 Years, 1997–2007

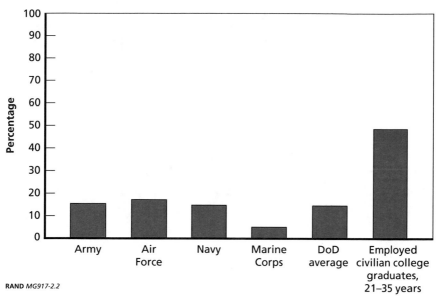

RAND *MG917-2.2*

occupations and are proscribed from serving in certain military occupations and positions.

Race/Ethnicity Representation

As mentioned in the introduction, prior to January 2003, race categories consisted of black, white, and other. Those who identified themselves as "Hispanic" were defined as such, without regard to race; thus, race totals included "Hispanic" as a separate category. From 2003 on, according to OMB guidelines, there were five race categories: American Indian or Alaska Native, Asian, black or African-American, Native Hawaiian or other Pacific Islander, and white. Respondents could also check more than one category, and this was reported as "two or more races." In addition, the percentage of respondents who elected not to respond (categorized as "unknown/elected not to respond") increased. Thus, the race percentages were calculated using all of these groups in the denominator (as is the case with DoD reporting). A separate question asked respondents their Hispanic identity; thus, race totals no longer included Hispanic as a category (OMB, 1997).

Figures 2.3 and 2.4 present the race/ethnicity composition of the officer corps by service in 1997–2002 and 2004–2007, respectively. No DoD race/ethnicity data for all officers were available for 2003. For comparison, the figures also show the race/ethnicity of civilian college graduates aged 21–49 years who were employed.

Prior to 2003, the majority of employed civilian college graduates were white (80 percent), as were the officers in the four services (79–86 percent). The Army had a higher proportion of blacks in its officer corps than was the case among employed civilian college graduates (11 percent versus 8 percent). About 6 percent of officers in the other three services were black. With the exception of the Air Force (2 percent), the proportion of Hispanics in the other three services was comparable to that among employed civilian graduates (4–5 percent).

Figure 2.3
Officer Corps, by Service, and Employed Civilian College Graduates, by Race/Ethnicity, 1997–2002

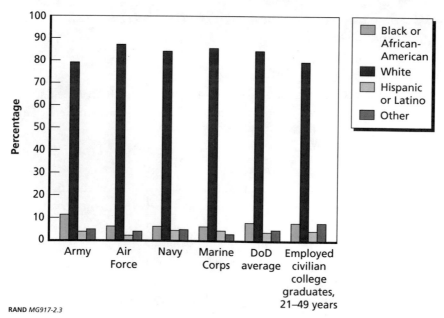

RAND *MG917-2.3*

Figure 2.4
Officer Corps, by Service, and Employed Civilian College Graduates, by Race/Ethnicity, 2004–2007

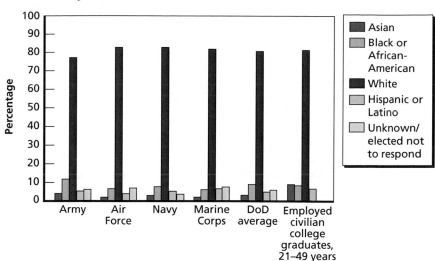

NOTE: Data were not provided on employed civilian college graduates whose race/ethnicity was unknown or who elected not to respond.

RAND MG917-2.4

About 8 percent of employed civilian college graduates were categorized as "other," as were 3–5 percent of officers (not shown).

We cannot compare race/ethnicity representation prior to and after 2003 because of the change in the way data on race were reported.

From 2004–2007, whites and blacks were equally represented among all DoD officers and the civilian college graduate workforce (81 percent and 9 percent, respectively). However, the percentage of black officers in the services ranged from 12 percent in the Army to 6 percent in the Marine Corps. Asians were underrepresented in the services, comprising 2–4 percent of military officers, compared with 9 percent of employed civilian graduates. Other groups (American Indians or Alaska Natives, Native Hawaiians or other Pacific Islanders, and those reporting two or more races) are not shown because of small sample sizes. About 6 percent of the officers elected not to respond. Thus, the overall percentage of nonwhites was 13 percent across the four services, compared with 19 percent in the civilian college graduate workforce.

The percentage of Hispanics was 7 percent among employed civilian graduates and 5 percent among all officers (ranging from 4 percent in the Air Force to 6 percent in the Marine Corps).

All Officer Accessions

Figure 2.5 shows the total number of officer accessions from 1997 through 2007. Officer accessions from all commissioning sources increased 37 percent from 1997 to 2002, when they peaked at 21,518. The number of accessions declined from 2004 to 2006, returning to 1999 levels. However, 2007 saw an increase of 7 percent over the previous year—from 16,486 to 17,713.

Not surprisingly, the Army accounts for the largest share of officer accessions. From 1999–2004, this ranged from 30 percent to 34 percent; in recent years, the percentage has risen sharply and was 43 percent in 2007. The total number of Navy officer accessions has fluctuated over time. For example, in 2001–2002, it rose to over 5,000,

Figure 2.5
Total Number of Officer Accessions, by Service, 1997–2007

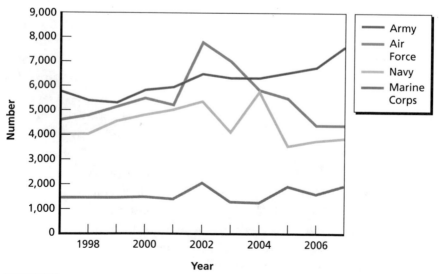

then dropped sharply to 4,123 in 2003. It rose again even more sharply to 5,700, only to decline to 3,500–3,800 in recent years. Over the most recent period, the Navy has accounted for 20–23 percent of all officer accessions. The Marine Corps accessed about 1,400–1,500 officers from 1997 to 2001. From then on, the number has fluctuated—up one year and down the next. In 2007, the number was close to 2,000. As a share of accessions, Marine Corps officers accounted for 10–11 percent of officers in more recent years. The number of Air Force officer accessions peaked in 2002 at 7,713, up sharply from the 4,600–5,500 in the five earlier years. The number has declined since then and was slightly less than 4,400 in 2007. Air Force officers accounted for 29–31 percent of accessions over most of this period (higher in 2002–2003), but this percentage has fallen in recent years and stood at 25 percent in 2007.

Gender Representation

Figure 2.6 shows the percentage of women among officer accessions by service and for all of DoD, averaged over the 11-year period. Compared with all officers, a higher proportion of officer accessions are women. For example, while 15 percent of Army and Navy officers, 5 percent of Marine Corps officers, and 17 percent of Air Force officers are women, the percentages of women officer accessions are 21 percent, 19 percent, 8 percent, and 23 percent, respectively, in the four services. The percentage of women officer accessions in the Navy and Air Force has increased slightly in recent years (1–2 percentage points). The percentage of women officer accessions in the Army and Marine Corps was highest in 2003–2004 (22 percent and 10 percent, respectively) and has declined slightly since then (20 percent and 8 percent, respectively, in 2007).

Race/Ethnicity Representation

Figure 2.7 shows the average racial/ethnic profile of officer accessions from 1997 to 2002, and Figure 2.8 shows the race/ethnicity of officer accessions, averaged over 2003–2007.[1] As noted earlier, because of the

---

[1] Although no 2003 race/ethnicity data were available for all officers, data were available for officer accessions for 2003.

Figure 2.6
Percentage of Women Among Officer Accessions, by Service, and Among Civilian College Graduates, 21–35 Years, 1997–2007

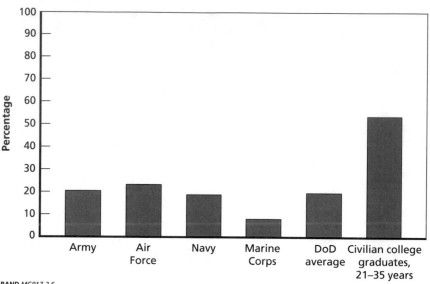

RAND *MG917-2.6*

change in the way race/ethnic data were collected and reported in 2003, we cannot directly compare the two figures. We also follow DoD in comparing the race/ethnicity representation of officer accessions to that of all civilian college graduates aged 21–35 years.

Overall, DoD officer accessions resembled civilian college graduates with respect to race/ethnicity during 1997–2002. For example, 78–79 percent of all officer accessions and all civilian college graduates during 1997–2002 were white, 8–9 percent were black, and 8–9 percent were "other." However, the percentage of whites and blacks varied across the services, with the Army having the highest percentage of nonwhites (25 percent). The percentage of black officer accessions ranged from 7 percent in the Marine Corps and Air Force to 12 percent in the Army. The percentage of Hispanics ranged from 2 percent in the Air Force to 7 percent in the Marine Corps over this period. With the exception of the Air Force, Hispanics were well represented in the other three services relative to the comparison group of college graduates during this period.

Figure 2.7
Officer Accessions, by Service, and Civilian College Graduates, 21–35 Years, by Race/Ethnicity, 1997–2002

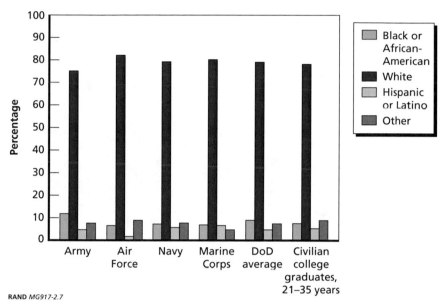

RAND *MG917-2.7*

Turning now to Figure 2.8, we see that the percentage of whites was somewhat lower among DoD officer accessions than among civilian college graduates (76 percent versus 79 percent) during 2003–2007. However, there is a fairly sizable percentage of "unknowns" among new officers—from 4 percent in the Navy to 16 percent in the Marine Corps.[2]

About 8 percent of DoD officer accessions and civilian graduates were black, although the percentage varied considerably by service (from 4 percent in the Marine Corps to 12 percent in the Army). Asians were underrepresented in the services—2–5 percent of officer accessions were Asian, compared with 11 percent of civilian college graduates. A small percentage checked "two or more races"—1 percent across DoD and among civilian graduates.

---

[2] "Unknowns" are included in the demonimator when calculating race percentages.

Figure 2.8
Officer Accessions, by Service, and Civilian College Graduates, 21–35 Years, by Race/Ethnicity, 2003–2007

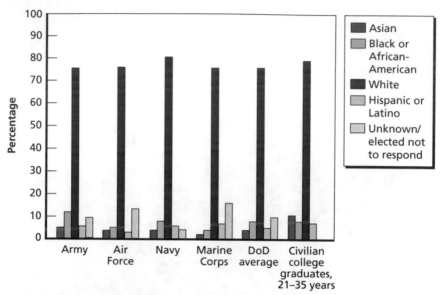

NOTE: Data were not provided on employed civilian college graduates whose race/ethnicity was unknown or who elected not to respond.

RAND *MG917-2.8*

About 6–7 percent of new officers were Hispanic, similar to their representation among civilian college graduates. The Air Force attracted a smaller percentage of Hispanics relative to the other services.

Overall, it seems that new officer accessions and all officers tend to be similar with respect to race/ethnicity. For example, while 10–16 percent of officers were nonwhite during 2004–2007, the percentage of officer accessions who were nonwhite ranged from 8 to 17 percent. However, the larger percentage of "unknowns" among officer accessions makes it difficult to draw definitive conclusions as to whether diversity is higher or lower among new officer accessions relative to all officers.

College Graduation Rates in Four-Year Institutions

The next three chapters present data on entering and graduating cohorts from the three service academies, including graduation rates of various entry cohorts by race/ethnicity and gender. To place these graduation rates in context, we examined data on six-year college graduation rates in four-year institutions, published by NCES. NCES maintains data on postsecondary institutions through the Integrated Postsecondary Education Data System (IPEDS).

> It is a system of interrelated surveys conducted annually by the [U.S. Department of Education's] National Center for Education Statistics (NCES). IPEDS gathers information from every college, university, and technical and vocational postsecondary institution that participates in the federal student financial aid programs. The Higher Education Act of 1965, as amended, requires that institutions that participate in federal student aid programs report data on enrollments, program completions, graduation rates, faculty and staff, finances, institutional prices, and student financial aid. . . .
>
> More than 6,700 institutions complete IPEDS surveys each year. These include research universities, state colleges and universities, private religious and liberal arts colleges, for-profit institutions, community and technical colleges, non-degree-granting institutions such as beauty colleges, and others. (NCES, undated[a])

The Graduation Rate Survey is a component of IPEDS and is fielded annually to collect data on the

> [n]umber of students entering the institution as full-time, first-time degree or certificate-seeking students in a particular year (cohort), by race/ethnicity and gender; [n]umber of students completing their program within a time period equal to one and a half times (150%) the normal period of time; and [n]umber of students who transferred to other institutions. (NCES, undated[b])

Horn (2006) analyzed 2004 graduation rates for a cohort of students who enrolled in 1998. Institutions were categorized by their 2000 Carnegie Classification (as primarily doctoral, master's, or bachelor's degree–granting institutions); by a measure of selectivity as very, moderately, or minimally selective (using a measure developed by Cunningham, 2005, that is based on several IPEDS variables, including college admission test scores, the number of applicants, and the number of students admitted); and by the size of their low-income enrollment (based on the proportion of federal grant aid recipients in the freshman cohort on which the graduation rates are based). The universe encompassed a total of 1,301 institutions, including 512 baccalaureate institutions, of which 117 were classified as "very selective." Enrollment-weighted six-year graduation rates were calculated for freshmen enrolled in the fall of 1998 who had never attended college before, attended full-time when they began college, and intended to earn a degree. Thus, the analysis does not take into account students who returned to college after dropping out, enrolled part-time, or enrolled in the spring. The average graduation rate for all four-year institutions was 57 percent.

Because the service academies tend to have highly selective admission criteria in terms of academic, physical, and leadership requirements, we believe that the best comparison is likely the "very selective" group of four-year institutions. We first provide some basic enrollment statistics on the 1998 entering freshman class. The percentage of women in these very selective four-year institutions was 57 percent, and the percentage of nonwhites was 23 percent. Overall, 8 percent of freshmen in very selective institutions were black, 3 percent were Hispanic, 5 percent were Asian, and less than 0.5 percent were American Indian. As we show later, while the percentage of women is considerably higher in these selective four-year institutions than in the service academies, the percentage of minorities is similar in the two types of institutions.

Overall, 76 percent of whites and 79 percent of Asians enrolled in these institutions graduated within six years, as did 71 percent of Hispanics. Only 60 percent of blacks and 62 percent of American Indians attending these colleges graduated. Women graduated at a higher rate than men (77 percent versus 73 percent). One point to note is that the data provided here are six-year graduation rates, while the majority of service academy entrants graduate in four years. However, it is likely that most students at very selective institutions also graduate within four years, making the comparison valid. As we show later, the graduation rates at the service academies are comparable to or higher than those in these very selective four-year colleges.

These data provide context for the academy graduation rates presented in the next three chapters. The next chapter examines data on USMA entrants and graduates.

CHAPTER THREE

Selected Diversity Rates and Trends: United States Military Academy

Chapters Three through Five are organized in a similar fashion. We begin with a note on terminology: Our data are organized by year of graduation, but it is sometimes easier to identify these classes by their year of entry into the academy. We assume that each class entered four years earlier than the year of graduation. This assumption may introduce a small amount of error (for example, if students had to drop out in a previous year and returned to join another class). However, because we are interested in broad trends over time and across categories, we believe this is a justifiable assumption. So, for example, we assume that those who graduated from USMA between 1996 and 2009 primarily entered between 1992 and 2005.

We received data from the service academies that described the makeup and outcomes of academy entrants who graduated in the previous 14 graduating classes (i.e., those who graduated between 1996 and 2009 and, in our terminology, entered between 1992 and 2005). In addition, as mentioned in Chapter Two, we also have data on classes that will graduate between 2010 and 2013 and so entered the academy between 2006 and 2009. Thus, we have a total of 18 entering classes. For this group, we examine (1) the gender and racial/ethnic makeup of the classes and (2) two outcomes of interest disaggregated by gender and race/ethnicity: the percentage of entrants who successfully completed the first year and entered the second year and the percentage who graduated from the academy.

From the DoD point of view, two longer-term outcomes are of interest. The first is whether individuals complete their ISO, and the second is whether they choose to stay in the military beyond that period. Obviously, the real payoff on the investment the military has made in these cadets/midshipmen comes when these officers choose the military as a career. To examine these questions, the service academies provided data for three earlier classes (those graduating between 1993 and 1995) with respect to the percentages of those who completed ISO and were still in service as of June 30, 2008. Thus, we switch gears and focus on graduating classes, rather than entering classes, and analyze those who graduated between 1993 and 2003—a total of 11 graduating classes. Note that, as of June 2008, the earliest graduating class (1993) was ten years beyond its ISO period (or 15 years beyond graduation), while the class of 2003 was just completing its ISO (five years beyond graduation). As mentioned, a limitation of these data is that, for each year beyond graduation, we can track the experience of only one graduating class in terms of continuation.

To compare trends over time, we present data averaged over the three earliest and three most recent cohorts for which we have the relevant data.

Gender and Race/Ethnicity of Entering Classes, 1992–2009

Table 3.1 presents an overview of the data in terms of minimum, maximum, and average size of the groups to provide context for the discussion that follows. We do not show disaggregated data for Native Hawaiian/other Pacific Islanders or those who self-identified as belonging to two or more races because these data were reported only since 2003. We also do not show data on "unknowns." While we include American Indians/Alaska Natives in the race/ethnicity profiles, we do not show disaggregated outcome data for this group because of small sample sizes.

Figures 3.1 and 3.2 present data on the gender and racial/ethnic makeup of the classes that entered USMA between 1992 and 2009. Table 3.2 displays the gender and racial/ethnic profile of the three earliest (1992–1994) and three most recent (2007–2009) classes.

Table 3.1
Overview of the Data: Means and Ranges, 1992–2009 Entering Classes, USMA

Characteristic	Average	Range (minimum, maximum)
Gender		
Men	1,032	(985, 1,116)
Women	183	(135, 224)
Race/ethnicity		
American Indian or Alaska Native	10	(4, 18)
Asian	71	(35, 99)
Black or African-American	81	(57, 113)
White	944	(844, 999)
Hispanic or Latino	80	(46, 120)
Total number of entrants	1,215	(1,126, 1,302)

Figure 3.1
Entering Classes by Gender, 1992–2009, USMA

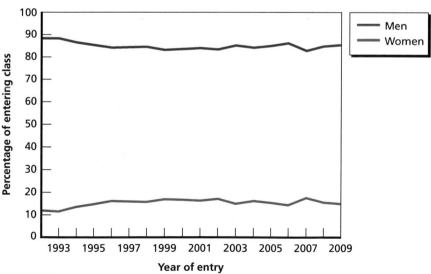

RAND *MG917-3.1*

Figure 3.2
Entering Classes by Race/Ethnicity, 1992–2009, USMA

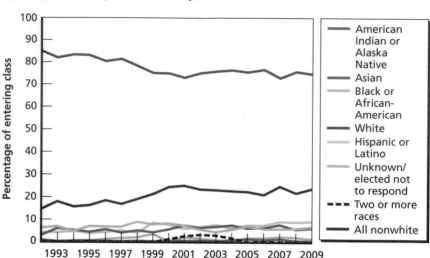

Over this period, the percentage of women in the USMA ranged from 11 percent to 17 percent, with an average of 15 percent. As Table 3.2 shows, the percentage of women increased from 12 percent in the earliest entry cohorts to 16 percent in the most recent entry cohorts.

To calculate race/ethnicity percentages, we followed DoD in including "unknown/elected not to respond" and "two or more races" in the denominator. Overall, whites accounted for between 73 percent and 85 percent of all academy entrants from 1992 to 2009 (see Figure 3.2).[1] The percentage of nonwhites (which includes those identifying themselves as "two or more races") has been trending upward over this period, and the entering classes of 2000, 2001, 2007, and 2009 had the highest proportion of nonwhites (24–25 percent). As shown in Table 3.2, the percentage of nonwhites was 16 percent in the 1992–1994 entering classes and 23 percent in the 2007–2009 entering

---

[1] The figure does not show the very small percentages categorized as Native Hawaiian or other Pacific Islander.

Table 3.2
Profile of Entering Classes, 1992–1994 and 2007–2009, USMA

Characteristic	1992–1994 Entering Classes (%)	2007–2009 Entering Classes (%)
Women	12	16
Nonwhite[a]	16	23
Race/ethnicity[b]		
American Indian or Alaska Native[c]	1	1
Asian	5	7
Black or African-American	6	6
White	83	75
Hispanic or Latino	4	9

[a] "Nonwhite" includes all race categories except "white" and "missing/refused to answer."

[b] The small percentages of Native Hawaiian/other Pacific Islanders, those indicating two or more races, and those with missing data or who refused to answer are not shown, but all are included in the total.

[c] Small sample sizes.

classes.[2] In the most recent cohorts, about 7 percent of entrants were Asian, 6 percent were black, and 9 percent were Hispanic.

First-Year Completion and Graduation Rates

For classes that entered USMA between 1992 and 2008, we calculated the rates of successful completion of the first year and entry into the second year and, for those who entered between 1992 and 2005, graduation from the academy. Figure 3.3 shows the average percentage of

---

[2]   As noted earlier, as of 2003, Hispanic ethnicity is collected separately from the race categories and is not included in the race totals. However, the academies were able to provide consistent data over time that included "Hispanic" as a separate, exclusive category within race/ethnicity. Despite this, it is not clear whether race/ethnicity profiles can be directly compared before and after 2003 because the availability of new race categories and a separate ethnic category may have led to changes in the way individuals identified themselves.

men and women who successfully completed the first year and entered the second year and the average percentage who graduated four years later, averaged over the three earliest and three most recent cohorts for which we have data. Figure 3.4 shows the variation in these two outcomes by gender and entering class.

On average, in the earliest entry cohorts, 80 percent of women and 85 percent of men successfully made it through the first year. The first-year completion rate improved over time: For the three most recent cohorts (2006–2008), 90 percent of women and 91 percent of men completed the first year and entered the second year.

Overall, the graduation rate for men remained stable over time—78 percent for the earliest cohorts and 77 percent for the most recent cohorts. Women increased their graduation rate by 5 percentage points over the same period, from 69 percent to 74 percent.

Figure 3.3
Percentage Who Entered Second Year, 1992–1994 and 2006–2008 Entering Classes, and Who Graduated from the Academy, 1992–1994 and 2003–2005 Entering Classes, by Gender, USMA

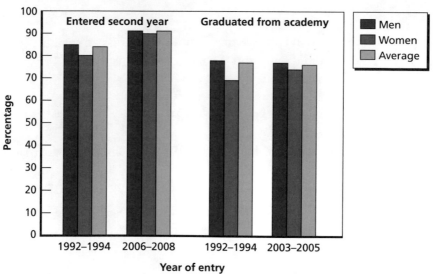

Figure 3.4
**Percentage Who Entered Second Year, 1992–2008, and Who Graduated
from the Academy, 1992–2005, by Gender and Entering Class, USMA**

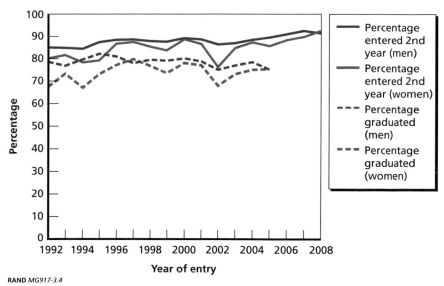

As illustrated in Figure 3.4, there has been a trend upward since 1995 in both outcomes, although outcomes for women tended to be more variable over time. In the 1998 and 1999 entering classes, there were declines in both the percentage of women who successfully completed the first year and the percentage of women who graduated from the academy. There was an even sharper decline in outcomes for women in the class of 2002, when the percentage who made it to the second year fell to 76 percent, and only 68 percent graduated. The most recent entering classes appeared to have rebounded, and outcomes were at or above average.

Figures 3.5 and 3.6 show the percentages completing the first year and graduating from the academy for different racial/ethnic groups, averaged over the three earliest and three most recent cohorts. First-year completion rates improved over time for every racial/ethnic group and averaged 91 percent in the most recent cohorts (2006–2008). For example, the first-year completion rate for Asians increased by 10 percentage points (from 85 percent to 95 percent) from the earliest to the

Figure 3.5
Percentage Who Entered Second Year, 1992–1994 and 2006–2008 Entering Classes, by Race/Ethnicity, USMA

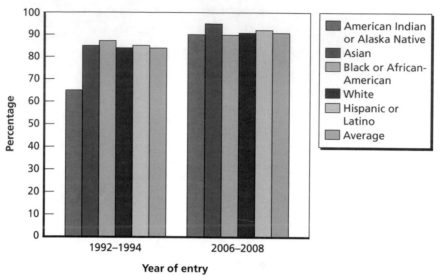

most recent entry cohorts. Similarly, Hispanics posted an improvement of 7 percentage points, as did whites. Blacks improved their completion rate by 3 percentage points. The number of American Indians/Alaska Natives was very small, but even this group's completion rate improved to 90 percent.

Figure 3.6 shows that graduation rates remained relatively stable between the earliest and most recent entry cohorts. For example, the graduation rate of blacks was 74 percent for both early and recent cohorts, and that of whites remained at 77 percent. Hispanics improved their graduation rate by 2 percentage points (74 percent to 76 percent), and Asians posted a surprising decline of 4 percentage points.

Figure 3.7 presents data on graduation rates of different racial/ethnic groups by entering class. Asians, who tended to have the highest graduation rates, showed a sharp decline for the entering classes of 2003 and 2004: Only 72 percent of the 2004 entering class graduated from the academy. The most recent year showed an increase. White cadets in the 2001 and 2002 entering classes experienced a small decline in

Figure 3.6
Percentage Who Graduated from the Academy, 1992–1994 and 2003–2005
Entering Classes, by Race/Ethnicity, USMA

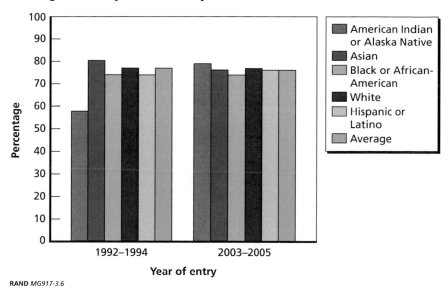

Figure 3.7
Percentage Who Graduated from the Academy, by Race/Ethnicity and
Entering Class, 1992–2005, USMA

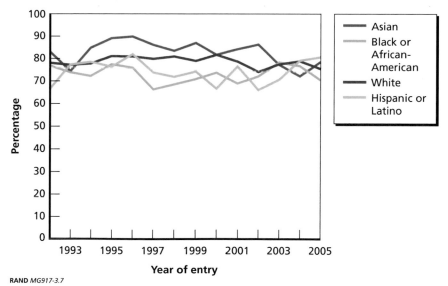

graduation rates; this reversed itself the following year but then declined in the most recent year. On the other hand, Hispanic cadets who entered in 2004 and 2005 had a markedly higher graduation rate (almost 14 percentage points higher) than those who entered in 2002 (80 percent versus 66 percent). In fact, they had the highest graduation rate of all racial/ethnic groups in the classes that entered in 2004 and 2005. With the exception of one year (2001), graduation rates of black cadets had been trending upward from 1997 to 2004 when they again turned down in 2005.

Conditional Graduation Rates, 1992–2005

The previous section examined outcomes as a percentage of the entering class. Another way to examine differences in graduation rates by gender or race/ethnicity is to look at conditional graduation rates, i.e., conditional on having successfully navigated the first-year hurdle. In other words, given that cadets successfully entered the second year, what is the likelihood of their successfully completing the remaining three years and graduating from the academy? Of necessity, conditional graduation rates will always be higher than unconditional graduation rates (because the denominator is smaller). However, if the difference between the two is fairly sizable, this would indicate that most attrition from the academy takes place during the first year and that reducing the level of attrition would raise overall graduation rates, provided that late attrition remains constant.[3]

Table 3.3 shows unconditional and conditional graduation rates by gender and race/ethnicity as well as the overall average for the earliest and most recent cohorts for which we have data. As shown in earlier figures, among the 1992–1994 entry cohorts, 77 percent graduated. However, the conditional graduation rate of those who made it to the second year was 92 percent (Table 3.3). Interestingly, although the first-year completion increased to 91 percent for the more recent cohorts,

---

[3] For example, consider a class of 100 students, 90 of whom make it to the second year and 80 of whom graduate. The unconditional rate is 80 percent while the conditional rate is 89 percent. The denominator for the conditional rate will always be smaller than for the unconditional rate, unless there is no attrition in the first year.

Table 3.3
Unconditional and Conditional Graduation Rates, 1992–1994 and
2003–2005 Entering Classes, by Gender and Race/Ethnicity, USMA

Characteristic	Unconditional Graduation Rate (%)		Conditional Graduation Rate (%)	
	1992–1994 Entering Classes	2003–2005 Entering Classes	1992–1994 Entering Classes	2003–2005 Entering Classes
Average	77	76	92	87
Gender				
Men	78	77	92	87
Women	69	74	86	87
Race/ethnicity[a]				
American Indian or Alaska Native[b]	58	79	88	92
Asian	80	76	93	85
Black or African-American	74	74	86	82
White	77	77	92	88
Hispanic or Latino	74	76	89	86

[a] The small percentage indicating two or more races is not shown.
[b] Small sample sizes.

the conditional graduation rate was somewhat lower than that of earlier cohorts—87 percent.

While women in the 2003–2005 entering classes had a slightly lower unconditional graduation rate than men (74 percent versus 77 percent), there was no difference in the conditional graduation rates of men and women in the most recent cohorts.

The conditional graduation rates for almost every racial/ethnic group declined from the earliest to the most recent cohorts. Provided cadets made it to the second year, their chances of graduation were quite high, although not similar, across all racial/ethnic groups (82–

88 percent for the most recent cohorts, with one exception).[4] Both blacks and Asians had lower-than-average conditional graduation rates in the most recent cohorts—surprising for Asians, who tended to have the highest graduation rates among all nonwhite groups generally.

Statistically Significant Differences in First-Year Completion and Graduation Rates

As mentioned in Chapter One, we tested for statistically significant differences in first-year completion and graduation rates among racial/ ethnic groups, men and women, and entry cohort years. For the sake of brevity, we refer to statistically significant differences as *significant differences*. The overall findings were as follows:

- Compared with white cadets, Asian cadets had significantly higher first-year completion and graduation rates. Blacks and Hispanics had significantly lower graduation rates than whites, although there was no significant difference in first-year completion rates between whites and either of these groups.
- Compared with men, women had significantly lower first-year completion and graduation rates.
- Compared with the entering class of 2008:
  - All entering classes from 1992 to 2005 had significantly lower first-year completion rates.
  - Classes that entered in 2006 and 2007 had first-year completion rates that were similar.
- Compared with the entering class of 2005:
  - Classes that entered in 1995, 1996, 1998, and 2000 had significantly higher graduation rates.
  - There were no significant differences in graduation rates between other entering classes and the entering class of 2005.

---

[4]   The conditional graduation rate for American Indians/Alaska Natives was 92 percent for the 2003–2005 cohort, but, as mentioned earlier, the sample size was quite small.

ISO Completion Rates of Graduates

As mentioned earlier, we have data on earlier graduating classes for which we can track both the decision to complete ISO and the decision to stay in the service beyond that point. We include data on the graduating class of 2003, although these officers were just completing their ISO in June 2008. One point to note is that we now show data by graduating class rather than by entering class because we are interested in the ISO completion rates of graduates, not entrants.

As before, we show data on the three earliest (1993–1995) and three most recent (2001–2003) graduating classes for which we have data. Figure 3.8 shows the average completion rates for the earliest and most recent graduating classes by gender, and Figure 3.9 shows ISO completion rates for all graduating classes by gender.

Among graduates, rates of ISO completion increased between the earliest and most recent cohorts by 9 percentage points (from 82 percent to 91 percent). Whereas earlier there was a 3-percentage-point

Figure 3.8
Percentage of Graduates Who Completed Their ISO, 1993–1995 and 2001–2003 Graduating Classes, by Gender, USMA

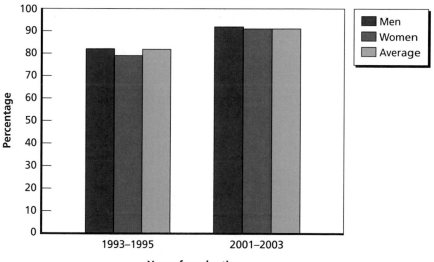

difference between men and women, there is little difference in the most recent cohorts (92 percent versus 91 percent).

With the exception of the earliest graduating class shown in Figure 3.9 (the 1993 graduating class), there is little difference in the rates at which men and women who graduated from USMA completed their ISO. For those in the earlier graduating class—1993—this represents 15 years beyond graduation and ten years beyond the initial obligation.

The ISO completion rates improved substantially across all racial/ethnic groups in the 2001–2003 graduating classes and ranged from 88 percent for blacks to 94 percent for Asians (Figure 3.10). The increase in ISO completion rates was 8 percentage points among blacks, 10 percentage points among Hispanics and whites, and 12 percentage points among Asians.

Figure 3.11 shows rates of ISO completion for the 1993–2003 graduating classes by race/ethnicity. Ignoring the earliest graduating classes, the figure shows that, on average, rates of ISO completion

Figure 3.9
Percentage of Graduates Who Completed Their ISO, by Gender and Graduating Class, 1993–2003, USMA

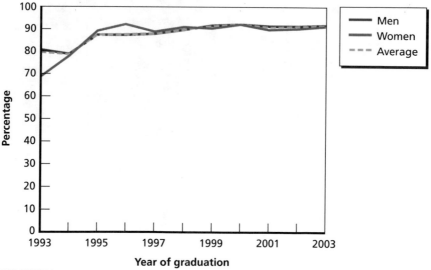

Figure 3.10
**Percentage of Graduates Who Completed Their ISO, 1993–1995 and
2001–2003 Graduating Classes, by Race/Ethnicity, USMA**

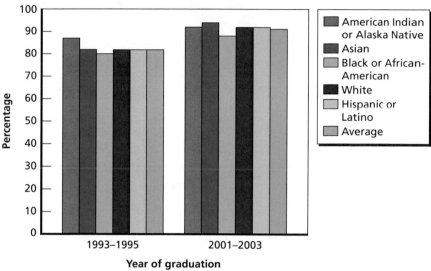

RAND *MG917-3.10*

Figure 3.11
**Percentage of Graduates Who Completed Their ISO, by Race/Ethnicity and
Graduating Class, 1993–2003, USMA**

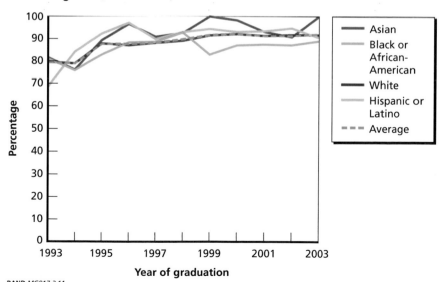

RAND *MG917-3.11*

were relatively high across most racial/ethnic groups (close to or above 90 percent). Black graduates tended to have somewhat lower rates of ISO completion, although in recent years, about 87 percent of black graduates completed their ISO, and black graduates in the graduating class of 2003 had an 89-percent completion rate.

Statistically Significant Differences in ISO Completion Rates

Overall, across all cohorts, the findings were as follows:[5]

- Asians had significantly higher ISO completion rates and blacks had significantly lower ISO rates than whites, although the latter had a wide confidence interval and the likely difference is small. There was no difference between Hispanics and whites with respect to ISO completion.
- There was no difference between the ISO completion rates of men and women.
- Those who graduated between 1993 and 1997 had significantly lower ISO completion rates than those who graduated in 2003.

Continuation in Service as of June 30, 2008

We now examine the continuation rates of those who graduated from USMA between 1993 and 2003. Like ISO completion rates, continuation in service can be affected by a host of factors—including policies to reduce end strength, promotion opportunities, attractive civilian job opportunities, warfare specialty, deployment, performance, and perceived discrimination, to name a few.

We calculated continuation rates in two ways—as a percentage of graduates from the academy and as a percentage of graduates who completed their ISO. We believe the first is important because it measures the return on investment in graduates. However, as mentioned earlier, ISO completion rates can be affected by exogenous factors—

---

[5] For the sake of brevity, we refer to statistically significant differences as *significant differences*.

for example, a policy to reduce end strength that may be outside the control of the graduates and that may result in separation. If this is the case, the first measure may understate the true continuation rate. Thus, while we show the percentage of graduates who continued in service as a function of the number of years beyond graduation in this section, Appendix A shows continuation rates for those who completed their ISO. We discuss both measures in the text, at least with respect to average continuation rates.

The figures in this section and in Appendix A show years since graduation and years beyond ISO completion, respectively, on the x-axis, so the most recent cohorts are on the left in the figures, and the earliest cohorts are on the right. Our data on the service academy graduates are current as of June 30, 2008. For the convenience of the reader, Table 3.4 shows a crosswalk between the year of graduation, the number of years since graduation (as of June 2008), and the number of years beyond ISO.

Table 3.4
Crosswalk Between Graduating Class Year, Years Since Graduation, and Years Beyond ISO, All Service Academies

Graduating Class (year)	Years Since Graduation (as of June 30, 2008)	Years Beyond ISO
1993	15	10
1994	14	9
1995	13	8
1996	12	7
1997	11	6
1998	10	5
1999	9	4
2000	8	3
2001	7	2
2002	6	1
2003	5	0

An important caveat is that each data point is provided by the experience of just one class. Our data do not allow us to examine whether the intermediate outcomes of each class were similar. For example, we do not know the percentage of any other earlier graduating class (say, the class of 2000) that remained in the first year beyond ISO (six years beyond graduation)—whether that percentage was 52 percent or higher or lower. All we can say is that if the experiences of these graduating classes are similar over time, then we would expect about 40 percent of graduates to stay at least three years beyond their ISO (eight years beyond graduation) and one-third of graduates to remain in the military for seven to ten years beyond their ISO (12–15 years beyond graduation).

Figure 3.12 shows the percentage of graduates who were still in service as of June 30, 2008, by gender and graduating class, and Figure A.1 in Appendix A provides similar data using the percentage of graduates who completed their ISO as the denominator. Because

Figure 3.12
Percentage of Graduates Remaining in Service as of June 2008, by Gender and Graduating Class, 1993–2003, USMA

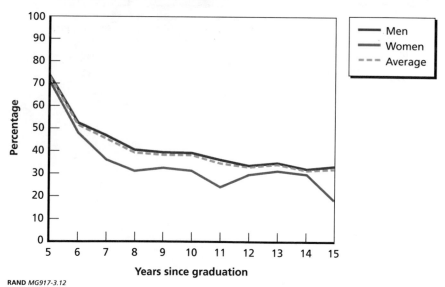

the ISO completion rates are high, there is not much difference in how continuation rates are calculated.

Looking first at the average, we found that, of the graduating class of 2003, only 73 percent of graduates (80 percent of those who had completed their ISO) remained in service as of June 2008, when they had just completed their ISO. For the graduating class of 2002, there was a sharp drop-off in the continuation rates of graduates—only 52 percent (57 percent of those who had completed their ISO) remained in service one year beyond their ISO. By the graduating class of 2000, we see a leveling off in terms of separation. About 38–39 percent of graduates (42–43 percent of those who had completed their ISO) remained in service eight to ten years beyond graduation, or three to five years beyond the ISO point. We found that, over time, 32–35 percent of graduates (40 percent of those who had completed their ISO) stayed 11–15 years beyond graduation, or six to ten years beyond their ISO point.

Women had much lower continuation rates than men, and this was true of every graduating class. Immediately after and one year beyond ISO completion, we see a 3- to 4-percentage-point difference in the continuation rates of women and men (71 percent versus 74 percent and 48 percent versus 52 percent, respectively). This increased to 9–11 percentage points for the graduating classes of 2001 and 2000—these classes were seven to eight years beyond graduation, or two to three years beyond their ISO decision point. At six years beyond the ISO decision point (11 years beyond graduation), only 24 percent of women graduates of the class of 1997 remained in service as of June 2008, compared with 36 percent of men. After that, there was a leveling off in the separation rates of men and women, and between 30 and 35 percent of both groups stayed in the military for seven to nine years beyond their ISO (12–14 years beyond graduation). By the tenth year beyond ISO completion (15 years beyond graduation), only 18 percent of women graduates of the class of 1993 remained in service in June 2008, compared with 34 percent of men.

Figure 3.13 presents continuation rates for the 1993–2003 graduating classes by race/ethnicity. The following were among the findings:

- Compared with white graduates, nonwhite graduates tended to continue at higher rates one to two years after completing their ISO.
- The continuation rates of nonwhite graduates tended to be more variable, but, in general, they appeared to continue at higher rates six to nine years beyond their ISO. This was especially true of Hispanic graduates. For example, 40–45 percent of Hispanic graduates remained in service six to nine years beyond their obligation period (or 11–15 years beyond graduation), compared with 32–35 percent of white graduates.
- At the tenth year beyond ISO (graduating class of 1993), or 15 years beyond graduation, only 23 percent of black and Hispanic graduates remained in service, compared with one-third of Asian and white graduates.

Figure 3.13
Percentage of Graduates Remaining in Service as of June 2008, by Race/Ethnicity and Graduating Class, 1993–2003, USMA

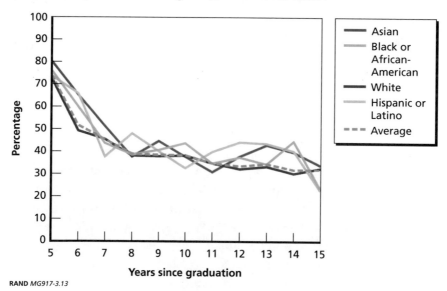

CHAPTER FOUR

Selected Diversity Rates and Trends: United States Air Force Academy

This chapter examines data on USAFA entrants and graduates and follows the previous chapter in terms of organization and presentation of data.

Gender and Race/Ethnicity of Entering Classes, 1992–2009

Table 4.1 presents an overview of the classes that entered USAFA between 1992 and 2009 in terms of gender and race/ethnicity. As mentioned in the previous chapter, the table does not include data on Native Hawaiian/other Pacific Islanders, those who self-identified as belonging to two or more races (because these data were only reported since 2003), and "unknowns."

Figures 4.1 and 4.2 show the breakdown by gender and race/ethnicity for all entering classes for which we have data, and Table 4.2 presents similar data on the three earliest and three most recent entry cohorts. Over this period, the percentage of women in the USAFA ranged from 13 percent to 21 percent, with an average of 17 percent. As Table 4.2 shows, the percentage of women increased from 15 percent in the 1992–1994 entry cohorts to 21 percent in the 2007–2009 entry cohorts.

Among the entering classes, whites accounted for between 75 percent and 84 percent of all academy entrants (see Figure 4.2).[1] Although

---

[1] The figure does not show the very small percentages categorized as Native Hawaiian or other Pacific Islander.

Table 4.1
Overview of the Data: Means and Ranges, 1992–2009 Entering Classes, USAFA

Characteristic	Average	Range (minimum, maximum)
Gender		
Men	1,050	(919, 1,140)
Women	223	(156, 289)
Race/ethnicity[a]		
American Indian or Alaska Native	18	(7, 34)
Asian	70	(39, 125)
Black or African-American	67	(41, 90)
White	1,021	(901, 1,112)
Hispanic or Latino	91	(70, 128)
Total number of entrants	1,273	(1,109, 1,383)

[a] The small percentage indicating two or more races is not shown.

Figure 4.1
Entering Classes, by Gender, 1992–2009, USAFA

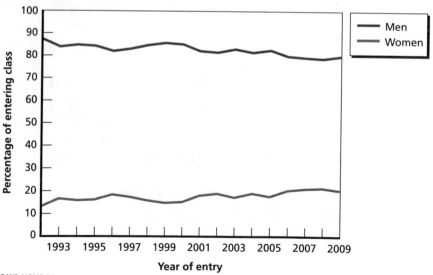

Figure 4.2
Entering Classes, by Race/Ethnicity, 1992–2009, USAFA

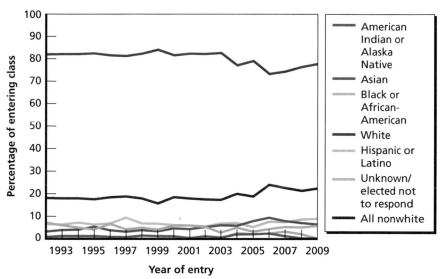

RAND *MG917-4.2*

the data provided by the academies on race/ethnicity appeared to be consistent, the data before and after 2003 may not be directly comparable because of the new reporting guidelines issued by OMB. However, there is a distinct upward trend in the percentage of nonwhites entering USAFA from 2004 on. In the most recent cohorts, about 22 percent of entrants were nonwhite—7 percent were Asian, 6 percent were black, 8 percent were Hispanic, and 1 percent were American Indian or Alaska Native.

First-Year Completion and Graduation Rates

For classes that entered USAFA between 1992 and 2007,[2] we have data on successful completion of the first year and entry into the second year and, for those who entered between 1992 and 2005, graduation

---

[2] While USMA and USNA provided data on the first-year completion rate of the 2008 entering class, USAFA did not. As a result, we show first-year completion for classes that entered between 1992 and 2007.

Table 4.2
Profile of Entering Classes, 1992–1994 and 2007–2009, USAFA

Characteristic	1992–1994 Entering Classes (%)	2007–2009 Entering Classes (%)
Women	15	21
Nonwhite[a]	18	22
Race/ethnicity[b]		
American Indian or Alaska Native[c]	1	1
Asian	4	7
Black or African-American	6	6
White	82	76
Hispanic or Latino	7	8

[a] "Nonwhite" includes all race categories except "white" and "missing/refused to answer."

[b] The small percentages of Native Hawaiian/other Pacific Islanders, those indicating two or more races, and those with missing data or who refused to answer are not shown, but all are included in the total.

[c] Small sample sizes.

from the academy. Figure 4.3 shows the average percentage of men and women who successfully completed the first year and entered the second year and who graduated four years later, averaged over the three earliest and three most recent cohorts for which we have data. Figure 4.4 shows the variation in these two outcomes by gender and entering class.

On average, about 81 percent of the 1992–1994 entering classes made it through the first year, and there was no difference between men and women in completion rates. Men improved their completion rate over time by 5 percentage points—86 percent of men in the 2005–2007 entering classes completed their first year, compared with 82 percent of women. As Figure 4.4 shows, outcomes for women have tended to be more variable over time and have generally decreased in recent years.

Figure 4.3
Percentage Who Entered Second Year, 1992–1994 and 2005–2007 Entering Classes, and Who Graduated from the Academy, 1992–1994 and 2003–2005 Entering Classes, by Gender, USAFA

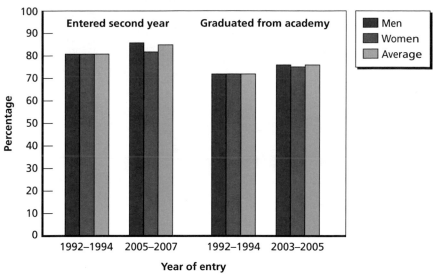

RAND *MG917-4.3*

About 72 percent of cadets in the 1992–1994 entering classes graduated from the academy. The graduation rate increased to 76 percent for the most recent cohorts, and there was little difference in the graduation rates of men and women.

Figures 4.5 and 4.6 show first-year completion and graduation rates for the various racial/ethnic groups, averaged over the earliest and most recent entry cohorts. The first-year completion rates of all racial/ethnic groups increased over time; the increases ranged from 2 percentage points for Hispanics to 8 percentage points for blacks. Whites had the lowest first-year completion rate among all racial/ethnic groups in the most recent cohorts (84 percent). Asians and blacks had the highest completion rates (91 percent and 89 percent, respectively). The completion rate for Hispanics was 86 percent. The sample sizes for American Indians/Alaska Natives were quite small, but this group posted a large improvement in completion rates across the two periods.

Figure 4.4
Percentage Who Entered Second Year, 1992–2007, and Who Graduated from the Academy, 1992–2005, by Gender and Entering Class, USAFA

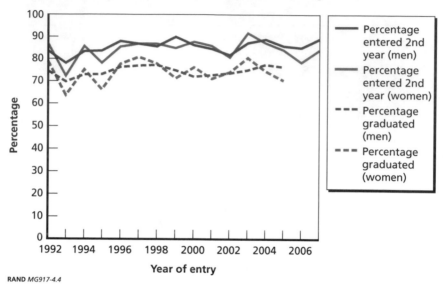

RAND *MG917-4.4*

Figure 4.5
Percentage Who Entered Second Year, 1992–1994 and 2005–2007 Entering Classes, by Race/Ethnicity, USAFA

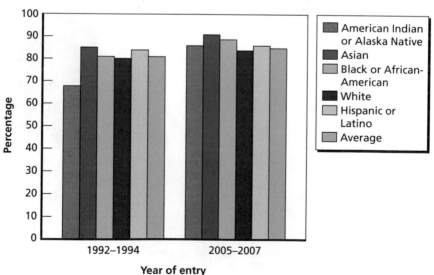

RAND *MG917-4.5*

Figure 4.6
Percentage Who Graduated from the Academy, 1992–1994 and 2003–2005
Entering Classes, by Race/Ethnicity, USAFA

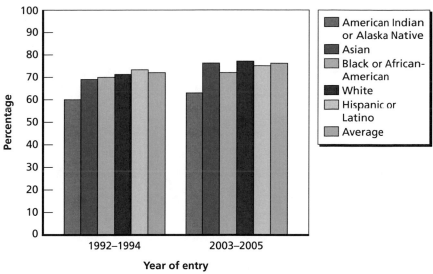

RAND *MG917-4.6*

Figure 4.6 shows that graduation rates also increased over time across all groups. Both Asians and whites experienced 6- to 7-percentage-point graduation rate increases over time, and their rates in the most recent cohorts were 76 percent and 77 percent, respectively. The graduation rate of Hispanics rose from 73 percent to 75 percent over the same period, and that of blacks rose from 70 percent to 72 percent.

While there is some variability in the percentage completing the first year by race/ethnicity, we focus on graduation rates because they show similar but somewhat larger trends in variability (see Figure 4.7). Asians, who tended to have the highest graduation rates, showed a sharp increase in graduation rates for the entering classes of 1996–1999 and a decline thereafter that reversed itself in 2003. Blacks had a large increase in rates in the 1997 entering class but then large drops in 2002 and 2005, while Hispanic cadets who entered in 1999 had a significant drop, followed by a gradual increase for subsequent entering classes.

Figure 4.7
**Percentage Who Graduated from the Academy, by Race/Ethnicity and
Entering Class, 1992–2005, USAFA**

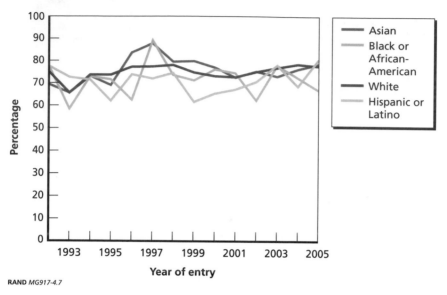

Conditional Graduation Rates, 1992–2005

As discussed earlier, we calculated conditional graduation rates for cadets who successfully made it through the first year.

Table 4.3 shows unconditional and conditional graduation rates by gender and race/ethnicity, as well as the overall average for the earliest and most recent cohorts for which we have data. As reported earlier, among the 1992–1994 entry cohorts, 72 percent graduated. However, the conditional graduation rate of those who made it to the second year was 89 percent. While the unconditional graduation rate increased by 4 percentage points, the graduation rate among those who successfully completed the first year declined slightly, to 87 percent. Women experienced a decline of 3 percentage points in the conditional graduation rate over this period.

As with the USMA results, the conditional graduation rates of every racial/ethnic group were higher than the unconditional graduation rates. For the 2003–2005 entry cohorts, the graduation rate of

Table 4.3
**Unconditional and Conditional Graduation Rates, 1992–1994 and
2003–2005 Entering Classes, by Gender and Race/Ethnicity, USAFA**

Characteristic	Unconditional Graduation Rates (%)		Conditional Graduation Rates (%)	
	1992–1994 Entering Classes	2003–2005 Entering Classes	1992–1994 Entering Classes	2003–2005 Entering Classes
Average	72	76	89	87
Gender				
Men	72	76	89	88
Women	72	75	89	86
Race/ethnicity[a]				
American Indian or Alaska Native[b]	60	63	87	81
Asian	69	76	81	86
Black or African-American	70	72	86	81
White	71	77	89	89
Hispanic or Latino	73	75	88	84

[a] The small percentage indicating two or more races is not shown.
[b] Small sample sizes.

those who successfully navigated the first year ranged from 81 percent
to 89 percent. The conditional graduation rates for blacks and Hispanics declined by 4–5 percentage points from the earliest to the most
recent cohorts, while that of whites remained the same and Asians
experienced an increase of 5 percentage points.

Statistically Significant Differences in First-Year Completion and Graduation Rates

As mentioned in Chapter One, we tested for significant differences
in first-year completion and graduation rates among racial/ethnic

groups, men and women, and entry cohort year. Overall, we found the following:[3]

- Asian cadets had significantly higher first-year completion rates than whites, and Hispanics had significantly lower graduation rates than whites. Other than that, there were no other significant differences between the first-year completion and graduation rates of the various racial/ethnic groups relative to whites.
- While women had a significantly lower first-year completion rate than men, the effect was small. There was no difference between men and women with respect to graduation from the academy.
- Compared with the entering class of 2007:
  - Classes entering in 1992–1995 and 2001–2002 had significantly lower first-year completion rates, although in some cases the effect was small.
- Compared with the entering class of 2005:
  - The 1993 and 1995 entering classes had significantly lower graduation rates.
  - All other classes had similar graduation rates.

ISO Completion Rates of Graduates

We can track the ISO completion rates of graduates for the 1993–2003 graduating classes. The officers in the graduating class of 2003 were just completing their ISO in June 2008.

Again, the figures in this and the next section show *graduating* classes rather than entering classes.

As before, we show data on the three earliest (1993–1995) and three most recent (2001–2003) graduating classes for which we have data. Figure 4.8 shows the average completion rates for these graduating classes by gender. The ISO completion rate declined among both men and women in the most recent cohorts—7 percentage points

---

[3] For the sake of brevity, we refer to statistically significant differences as *significant differences*.

Figure 4.8
Percentage of Graduates Who Completed Their ISO, 1993–1995 and 2001–2003 Graduating Classes, by Gender, USAFA

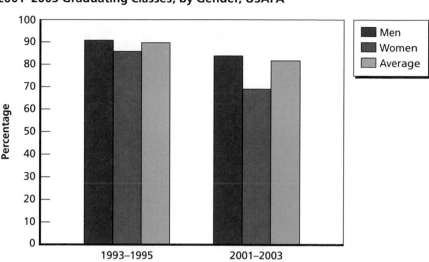

RAND *MG917-4.8*

among men and 17 percentage points among women. Exogenous factors—such as reductions in force or the civilian economy—are likely to affect retention and ISO completion and may help explain the decline. An analysis of the effects of such factors was beyond the scope of this study.

In Figure 4.9, the most recent graduating classes are on the left. Compared with men, women graduates exhibited greater variability in the rates at which they completed their ISO. As in Figure 4.8, recent cohorts showed a decrease in terms of the overall percentages of graduates who completed their ISO.

Figures 4.10 and 4.11 present data on ISO completion rates by race/ethnicity for the earliest and most recent graduating cohorts and by graduating class, respectively. We found that ISO completion rates declined for every racial/ethnic group between 1993–1995 and 2001–2003. These declines ranged from 4 percentage points for Hispanics to 14 percentage points for blacks. As mentioned earlier, these declines may be driven by exogenous factors—service policies to reduce strength, or

Figure 4.9
Percentage of Graduates Who Completed Their ISO, by Gender and Graduating Class, 1993–2003, USAFA

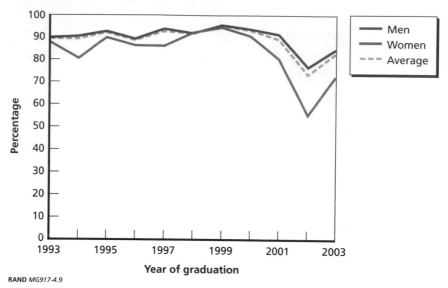

RAND *MG917-4.9*

Figure 4.10
Percentage of Graduates Who Completed Their ISO, 1993–1995 and 2001–2003 Graduating Classes, by Race/Ethnicity, USAFA

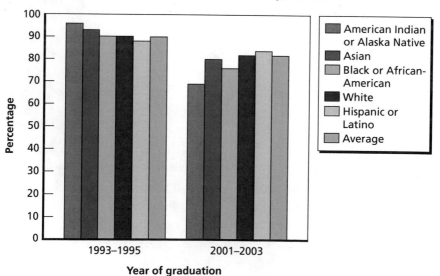

RAND *MG917-4.10*

Figure 4.11
Percentage of Graduates Who Completed Their ISO, by Race/Ethnicity and Graduating Class, 1993–2003, USAFA

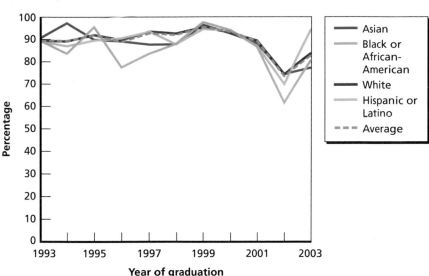

RAND *MG917-4.11*

competition from the civilian economy—that were beyond the scope of our study.

Figure 4.11 shows rates of ISO completion for the 1993–2003 graduating classes by race/ethnicity. Ignoring the most recent graduating classes, we found that rates of ISO completion were, on average, relatively high across most racial/ethnic groups (close to or above 85 percent). Black graduates tended to have somewhat lower rates of ISO completion in some years, and, except for Hispanics in the most recent year, all groups had lower ISO completion rates in the most recent two years.

Statistically Significant Differences in ISO Completion Rates
Across all cohorts, our analysis revealed the following:[4]

---

[4] For the sake of brevity, we refer to statistically significant differences as *significant differences*.

- Blacks had a significantly lower rate of ISO completion than whites, but there were no differences in ISO completion rates between whites and Asians or between whites and Hispanics.
- Women had significantly lower rates of ISO completion than men.
- Almost all earlier graduating classes (those that graduated between 1993 and 2001) had significantly higher ISO completion rates than the graduating class of 2003. The graduating class of 2002 had a significantly lower rate of ISO completion than the 2003 class.

Continuation in Service as of June 30, 2008

We now examine the percentage of graduates who remained in service after graduation. As in the previous section, the data shown are for graduating classes, but we show the number of years since graduation on the x-axis rather than year of graduation. Thus, the order is reversed, so data for the most recent graduating class is on the left in the figure. As a reminder, Table 3.4 in Chapter Three shows a crosswalk between the graduating classes, years since graduation, and years beyond ISO used in the figures. It is important to remember that many factors affect continuation in service, including reductions in force, warfare specialty, promotion opportunities, civilian opportunities, and perceived discrimination, among others. Examining the reasons for the trends shown here was beyond the scope of this study.

Figures 4.12 and 4.13 show the percentage of graduates who were still in service as of June 30, 2008, by graduating class and gender and by graduating class and race/ethnicity, respectively. Figures A.3 and A.4 in Appendix A present similar data for graduates who completed their ISO (instead of for all graduates). Because the ISO completion rates are high, there is little difference between the two sets of continuation rates.

Looking first at average continuation rates, 82 percent of officers in the graduating class of 2003 were still in service as of June 2008. If we restrict this sample to officers who had completed their ISO, we find that 99 percent were still in service in June 2008 (see Figure A.3).

Figure 4.12
Percentage of Graduates Remaining in Service as of June 2008, by Gender and Graduating Class, 1993–2003, USAFA

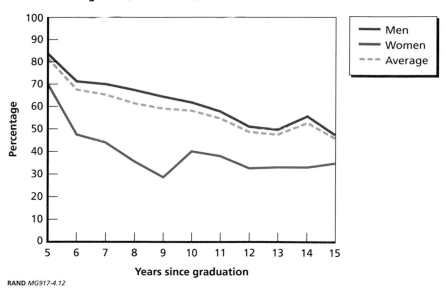

RAND *MG917-4.12*

Figure 4.13
Percentage of Graduates Remaining in Service, by Race/Ethnicity and Graduating Class, 1993–2003, USAFA

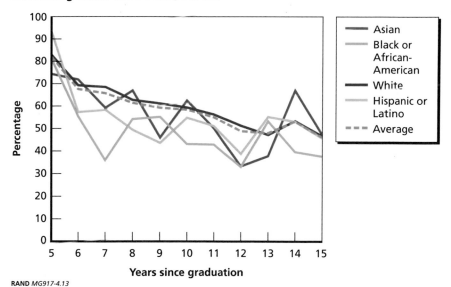

RAND *MG917-4.13*

For the graduating class of 2002, there was a slight drop-off in the continuation rates of graduates, with 67 percent remaining in service one year beyond their ISO. About 59–65 percent of graduates (or 62–66 percent of those who had completed their ISO) remained in service three to five years beyond their ISO. We found that, over time, 46–58 percent of graduates (52–60 percent of those who had completed their ISO) stayed six to ten years beyond their ISO.[5]

Again, we remind the reader of an important limitation of these data. Each data point reflects the experience of just one class. Our data do not allow us to examine whether the intermediate outcomes of each class are similar. For example, we do not know the percentage of any other earlier graduating class (say, the class of 2000) that remained in the first year beyond ISO (six years beyond graduation)—whether that 67 percent was higher or lower. All we can say is that if the experiences of these graduating classes are similar over time, we would expect about 60 percent of graduates to stay in service for at least three years beyond their ISO (eight years beyond graduation) and about one-half of graduates to remain in the military for seven to ten years beyond their ISO (12–15 years beyond graduation).

Women had much lower continuation rates than men, and this was true of every graduating class. Immediately after ISO completion and one year beyond ISO, we see a 14- to 24-percentage-point difference in the continuation rates of women and men (70 percent versus 84 percent and 47 percent versus 71 percent, respectively). This increased to 26–32 percentage points for the graduating classes of 2001 and 2000. These classes were two to three years beyond their ISO decision point. The largest difference was in the fourth year beyond the ISO decision point, in which only 28 percent of women graduates in the class of 1999 remained in service as of June 2008, compared with 64 percent of men. After that, differences remained in separation rates of men and women. By the tenth year after ISO completion (15 years

---

[5] This rate is higher than in USMA. A significant proportion of USAFA graduates become pilots and incur an additional service obligation of ten years of service after completing pilot training. The same is true for USNA graduates. See Appendix B for more information about service obligations.

beyond graduation), only 35 percent of women graduates in the class of 1993 remained in service in June 2008, compared with 47 percent of men.

Figure 4.13 presents continuation rates for graduating classes by race/ethnicity. Among the findings were the following:

- There is variability among racial/ethnic groups in terms of continuation, with Hispanic, white, and Asian graduates having higher continuation rates in certain years.
- Blacks in most years continued at lower rates than other groups.
- At the 15th year beyond ISO completion (graduating class of 1993), only 38 percent of black graduates remained in service, compared with 46–47 percent of other groups.

CHAPTER FIVE

Selected Diversity Rates and Trends: United States Naval Academy

This chapter follows the previous two with respect to organization and presentation and discussion of data. However, data provided by USNA allowed us to separate those who joined the Navy on graduation from those who joined the Marine Corps, and we do so in the last section of this chapter.

Gender and Race/Ethnicity of Entering Classes, 1992–2009

Table 5.1 presents an overview of the USNA entering classes of 1992 to 2009 to provide context for the discussion that follows. Once again, we exclude Native Hawaiian/other Pacific Islanders and those who self-identified as belonging to two or more races because these data were reported only since 2003. While we include American Indians/ Alaska Natives in the race/ethnicity profiles and for some outcomes, the sample sizes were very small.

Figures 5.1 and 5.2 show the profile of entrants by entering class and gender and race/ethnicity, respectively. Table 5.2 displays the gender and racial/ethnic profile of the 1992–1994 and 2007–2009 entry cohorts (the three earliest and most recent cohorts for which we have data). Between 1992 and 2009, the percentage of women in USNA ranged from 14 percent to 22 percent, with an average of 18 percent. Comparing the earliest to the most recent cohorts (see Table 5.2), we see that the percentage of women has increased from 15 percent to 21 percent over time.

Table 5.1
Overview of the Data: Means and Ranges, 1992–2009 Entering Classes,
USNA

Characteristic	Average	Range (minimum, maximum)
Gender		
Men	996	(942, 1,060)
Women	212	(166, 271)
Race/ethnicity[a]		
American Indian or Alaska Native	13	(4, 13)
Asian	46	(26, 59)
Black or African-American	76	(41, 100)
White	941	(789, 1,005)
Hispanic or Latino	106	(67, 178)
Unknown/elected not to respond	22	(42, 104)
Total number of entrants	1,208	(1,113, 1,228)

[a] The small percentage indicating two or more races is not shown.

Figure 5.2 presents the racial/ethnic makeup of the entering class-es.[1] Overall, whites accounted for between 72 percent and 84 percent of all academy entrants, with the exception of the 2009 entering class, when the percentage of whites fell to 64 percent. The percentage of nonwhites fell in recent years from 25 percent in 2002–2003 to 18 percent in 2007 but rebounded in 2008. In 2009, the percentage of nonwhites rose sharply to 28 percent.

On average, as Table 5.2 shows, 22 percent of the 2007–2009 entering classes were nonwhite. In these recent cohorts, about 4 percent of academy entrants were Asian, 5 percent were black, 12 percent were Hispanic, and less than 1 percent were American Indian or Alaska Native.

---

[1]   The figure does not show the percentages of Native Hawaiians or other Pacific Islanders.

Figure 5.1
Entering Classes, by Gender, 1992–2009, USNA

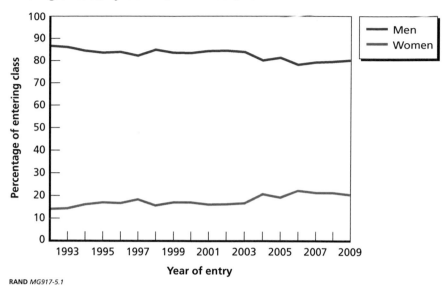

RAND *MG917-5.1*

Figure 5.2
Entering Classes, by Race/Ethnicity, 1992–2009, USNA

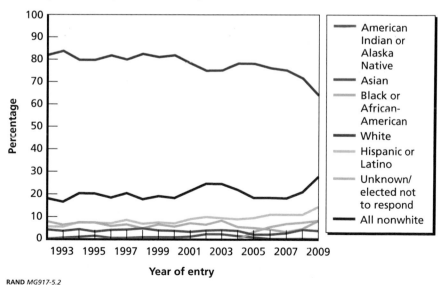

RAND *MG917-5.2*

Table 5.2
Profile of Entering Classes, 1992–1994 and 2007–2009, USNA

Characteristic	1992–1994 Entering Classes (%)	2007–2009 Entering Classes (%)
Women	15	21
Nonwhite[a]	18	22
Race/ethnicity[b]		
American Indian or Alaska Native[c]	1	< 1
Asian	4	4
Black or African-American	7	5
White	82	70
Hispanic or Latino	6	12

[a] "Nonwhite" includes all race categories except "white" and "missing/refused to answer."

[b] The small percentages of Native Hawaiian/other Pacific Islanders, those indicating two or more races, and those with missing data or who refused to answer are not shown, but all are included in the total.

[c] Small sample sizes.

First-Year Completion and Graduation Rates

Again, we calculated the percentage of entrants who successfully made it to the second year for those entering between 1992 and 2008. For those entering between 1992 and 2005, we also calculated the percentage of those who graduated from USNA.

Figure 5.3 shows the average percentage of men and women who successfully completed the first year and entered the second year and the average percentage who graduated four years later for the three earliest and three most recent cohorts for which we have data. Figure 5.4 shows the variation in these two outcomes by gender and entering class.

On average, about 83 percent of women and 89 percent of men in the 1992–1994 entry cohorts completed the first year and went on to the second year. Both groups improved their completion rates

Figure 5.3
Percentage Who Entered Second Year, 1992–1994 and 2006–2008 Entering Classes, and Who Graduated from the Academy, 1992–1994 and 2003–2005 Entering Classes, by Gender, USNA

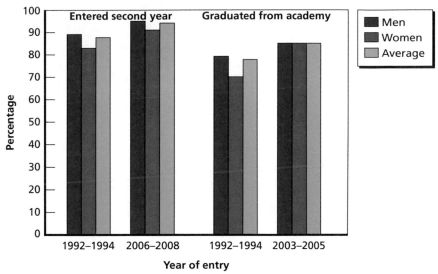

over time: The rate for women in the most recent cohorts increased to 91 percent (an 8-percentage-point gain), and the rate for men increased to 95 percent (a 6-percentage-point gain), reducing the gap between the completion rates of men and women to 4 percentage points.

The overall graduation rate improved substantially over this period as well, from 78 percent to 85 percent. Men experienced a 6-percentage-point improvement in graduation rates (from 79 percent to 85 percent), while women posted a substantial gain of 15 percentage points. Thus, whereas earlier, there was a sizable 9-percentage-point difference in the graduation rates of men and women, that difference was eliminated in the most recent cohorts, and the graduation rate was 85 percent for both in the 2003–2005 entry cohorts.

As Figure 5.4 shows, the improvement in both outcomes for women started in the late 1990s and became more marked from 2000 on. For example, the first-year completion rate for women in the entering classes after 2003 was above 90 percent. Similarly, the graduation

Figure 5.4
Percentage Who Entered Second Year, 1992–2008, and Who Graduated from the Academy, 1992–2005, by Gender and Entering Class, USNA

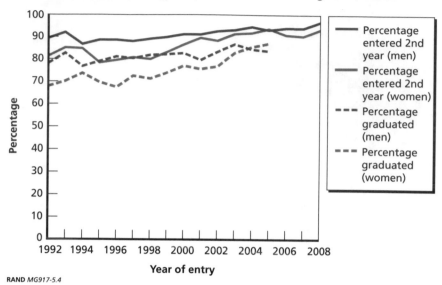

RAND *MG917-5.4*

rate among women in the entering classes from 2003 onward increased to above 80 percent for the first time. For two of the most recent classes, the graduation rate for women was higher than that of men— for example, 87 percent versus 84 percent in 2005.

Figures 5.5 and 5.6 show first-year completion and graduation rates for the various racial/ethnic groups averaged across the three earliest and three most recent cohorts. Again, we have small sample sizes for American Indians/Alaska Natives, so while we include them in Figure 5.5, we do not show disaggregated data by year for this group. Excluding this group, Asians had the highest first-year completion rate (90 percent) of all racial/ethnic groups in the earlier cohorts, although all the other groups had very high first-year completion rates as well (86–88 percent). First-year completion rates improved across all groups and stood at 93–94 percent for the cohorts entering between 2006 and 2008.

There were more marked differences in the graduation rates of the various racial/ethnic groups of students (see Figure 5.6). In the earliest cohorts, both Hispanics and blacks had much lower graduation

Figure 5.5
Percentage Who Entered Second Year, 1992–1994 and 2006–2008 Entering Classes, by Race/Ethnicity, USNA

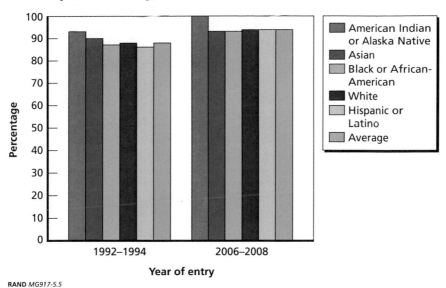

RAND *MG917-5.5*

Figure 5.6
Percentage Who Graduated from the Academy, 1992–1994 and 2003–2005 Entering Classes, by Race/Ethnicity, USNA

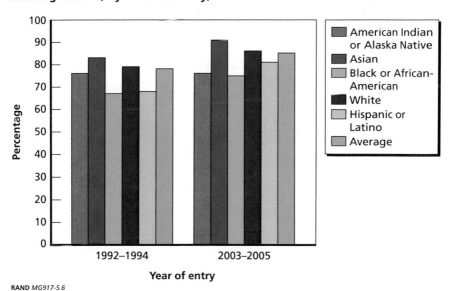

RAND *MG917-5.6*

rates (67–68 percent) than whites and Asians (79 percent and 83 percent, respectively). Every group improved its graduation rate over time. The graduation rate of whites and Asians increased by 7–8 percentage points to 86 percent and 91 percent, respectively, in the most recent cohorts. While the graduation rate for blacks increased by 8 percentage points, this group had the lowest graduation rate among all groups (75 percent). Hispanics showed the largest improvement in graduation rates—13 percentage points, from 68 percent to 81 percent.

Figure 5.7 shows year-by-year trends in graduation rates by race/ethnicity. Overall, there is variability in the rates, and, as seen in Figure 5.6, several groups showed an improvement. Asians, who tended to have the highest graduation rates, showed some surprisingly large ups and downs, but this may be the result of small sample sizes (on average, the entering class had about 40–50 Asian midshipmen). In recent years, the graduation rate increased from 72 percent for the 2001 entering class to 96 percent for the class of 2005. Blacks in

Figure 5.7
Percentage Who Graduated from the Academy, by Race/Ethnicity and Entering Class, 1992–2005, USNA

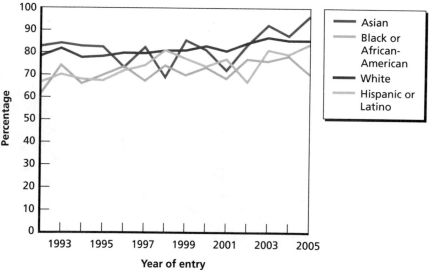

some recent entering classes (2002–2004) posted a marked improvement in graduation rates—from 69 percent the year prior (2001) to 76–78 percent. However, this rate fell to 71 percent in 2005. Hispanic midshipmen who entered between 2000 and 2005 experienced a sharp increase in graduation rates, with 83 percent graduating in the most recent year. Graduation rates for whites also showed a gradual upward trend, with the three most recent entering classes (2003–2005) graduating at a rate of 85–87 percent.

Conditional Graduation Rates, 1992–2005

As before for the USMA and USAFA cohorts, we calculated conditional graduation rates for USNA midshipmen who successfully made it through the first year and entered the second year. This rate contrasts with the unconditional graduation rate shown in Figure 5.7, which is calculated as a percentage of entrants. As noted earlier, these conditional graduation rates are of necessity higher than the unconditional graduation rates, but the magnitude of the difference between the two is an indicator of the relative extent of first-year versus later-year attrition.

Table 5.3 shows unconditional and conditional graduation rates by gender and race/ethnicity, as well as the overall average for the earliest and most recent cohorts for which we have data. As reported earlier, among the 1992–1994 entry cohorts, 78 percent graduated. However, the conditional graduation rate of those who made it to the second year was 88 percent. While the unconditional graduation rate increased by 7 percentage points, the graduation rate among those who successfully completed the first year also increased over time, to 91 percent. Women experienced a small decrease and men a small increase (2 percentage points) in the conditional graduation rate over this period.

For the 1992–1994 entry cohorts, the graduation rate of those who successfully navigated the first year ranged from 77 percent to 93 percent. The conditional graduation rates for blacks and Hispanics increased sharply by 8 and 10 percentage points, respectively, from the earliest to the most recent cohorts. The conditional graduation rates of other groups showed much smaller increases or remained stable over the same period.

Table 5.3
Unconditional and Conditional Graduation Rates, 1992–1994 and
2003–2005 Entering Classes, by Gender and Race/Ethnicity, USNA

Characteristic	Unconditional Graduation Rates (%)		Conditional Graduation Rates (%)	
	1992–1994 Entering Classes	2003–2005 Entering Classes	1992–1994 Entering Classes	2003–2005 Entering Classes
Average	78	85	88	91
Gender				
Men	79	85	89	91
Women	70	85	94	92
Race/ethnicity[a]				
American Indian or Alaska Native[b]	76	76	81	83
Asian	83	91	93	93
Black or African-American	67	75	77	85
White	79	86	90	92
Hispanic or Latino	68	81	79	89

[a] The small percentage indicating two or more races is not shown.
[b] Small sample sizes.

Statistically Significant Differences in First-Year Completion and Graduation Rates

As mentioned in Chapter One, we tested for significant differences in first-year completion and graduation rates by racial/ethnic group, gender, and entry cohort year. Across all years, we found the following:[2]

- There were no significant differences in first-year completion rates between whites and blacks or Asians. Hispanics had significantly

---

[2] For the sake of brevity, we refer to statistically significant differences as *significant differences.*

lower first-year completion rates than whites, but the effect was modest. Both blacks and Hispanics had significantly lower graduation rates than whites.

- Women had significantly lower first-year completion and graduation rates than men.
- Compared with the entering class of 2008:
  - All entering classes prior to 2004 had lower first-year completion rates.
  - Among the more recent entering classes, some classes had lower first-year completion rates but also had wide confidence intervals around the estimate, suggesting that the effects were weak.
- Compared with the entering class of 2005:
  - All classes that entered prior to 2000 had significantly lower graduation rates.
  - Among the more recent entering classes, the 2001 and 2003 entering classes also had significantly lower graduation rates.

ISO Completion Rates of Graduates

This section and the next discuss our findings in terms of graduating classes, rather than entering classes. We can track ISO completion rates for those who graduated between 1993 and 2003. The officers in the 2003 graduating class were just completing their ISO in June 2008.

As in Chapters Three and Four, we show data on ISO completion for the three earliest (1993–1995) and three most recent graduating classes (2001–2003) for which we have data. Figure 5.8 shows average completion rates by gender for these cohorts, and Figure 5.9 shows year-by-year trends in ISO completion by gender.

ISO completion rates have declined over time—from a high of 95 percent for the earliest cohorts to 89 percent for the most recent cohorts. As mentioned before, ISO completion rates are affected by a number of exogenous factors, some outside the control of the graduates, such as service policies to reduce end strength. The decline in the

Figure 5.8
Percentage of Graduates Who Completed Their ISO, 1993–1995 and 2001–2003 Graduating Classes, by Gender, USNA

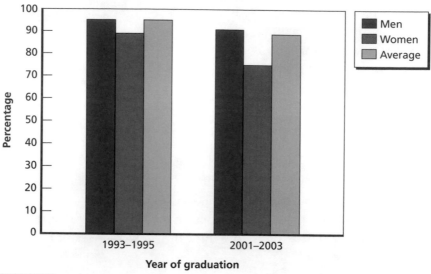

RAND *MG917-5.8*

ISO completion rate of women was particularly marked—from 89 percent to 75 percent for the 2001–2003 graduating classes, a decline of 14 percentage points. This resulted in a 16-percentage-point difference between the ISO completion rates of men and women in recent years, whereas before, the difference was only 6 percentage points.

Figure 5.9 shows that, with the exception of the most recent graduating class (2003), 90–97 percent of graduates completed their ISO. However, the percentage has been trending downward from a high of 97 percent for the class of 1998. Women's rates have fallen more sharply than those of men since that time and, as shown in Figure 5.8, have been particularly low in the three most recent cohorts.

Figure 5.10 shows ISO completion rates by race/ethnicity for the 1992–1994 and 2001–2003 graduating classes. In the earliest cohorts, the ISO completion rates of all racial/ethnic groups (with the exception of American Indians/Alaska Natives) were close to 95 percent, and there was little difference among the groups. In the most recent cohorts, the ISO completion rates of Asians and blacks were markedly lower (decreases of 13 and 11 percentage points, respectively) and

Figure 5.9
Percentage of Graduates Who Completed Their ISO, by Gender and
Graduating Class, 1993–2003, USNA

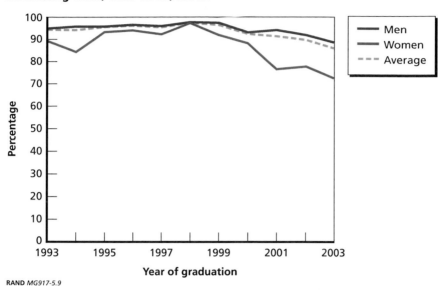

RAND *MG917-5.9*

Figure 5.10
Percentage of Graduates Who Completed Their ISO, 1993–1995 and
2001–2003 Graduating Classes, by Race/Ethnicity, USNA

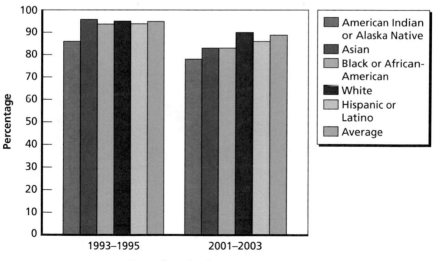

RAND *MG917-5.10*

well below average. Hispanics experienced a smaller decline (8 percentage points), but their ISO completion rate was also below average; the completion rate for whites declined 5 percentage points to 90 percent.

Figure 5.11 shows rates of ISO completion for the graduating classes of 1993–2003 by race/ethnicity. In general, the ISO completion rates of all groups were higher for the earlier graduating classes than the more recent classes: Almost 90 percent or more of graduates in these earlier classes completed their ISO, regardless of racial/ethnic group. As noted earlier, the decline in ISO completion rates started in 1999 for almost every group. In the most recent class, whites had an 88-percent ISO completion rate, but the completion rates for other groups were considerably lower—75 percent among Asian graduates, 79 percent among black graduates, and 74 percent among Hispanic graduates.

Figure 5.11
Percentage of Graduates Who Completed Their ISO, by Race/Ethnicity and Graduating Class, 1993–2003, USNA

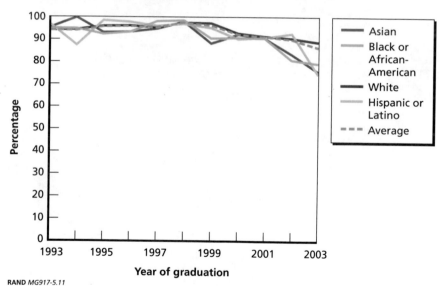

Statistically Significant Differences in ISO Completion Rates

Across all years, we found the following:[3]

- Asians, blacks, and Hispanics had significantly lower ISO completion rates than whites, but the effects were weak for Asians and Hispanics.
- Women had a significantly lower rate of ISO completion than men.
- All earlier graduating classes (1993–2002) had significantly higher ISO completion rates than the graduating class of 2003.

Continuation in Service as of June 30, 2008

This section examines the percentage of graduates who remained in service after graduation as of June 30, 2008. As in the previous section, the data shown are for graduating classes. However, we show the number of years since graduation on the x-axis, rather than year of graduation, so the order is reversed, with the most recent graduating class shown on the left (five years since graduation, or the graduating class of 2003). As a reminder, Table 3.4 in Chapter Three shows a crosswalk between the graduating classes, number of years since graduation, and years beyond ISO used in the figures. As before, it is important to keep in mind that several factors affect continuation in service—reductions in force, warfare specialty, promotion opportunities, civilian opportunities, and perceived discrimination, among others.[4] Examining the reasons for the trends we see here was beyond the scope of this study.

---

[3] For the sake of brevity, we refer to statistically significant differences as *significant differences*.

[4] In supplemental comments provided for this research, the Navy mentioned the importance of mentoring, for both retention at the service academies and postcommissioning, and identified the key role played by postcommissioning mentoring organizations, such as the National Naval Officers Association and the Association of Naval Service Officers, and their impact on minority retention.

Figures 5.12 and 5.13 show the percentage of graduates who were still in service as of June 30, 2008, by graduating class and gender and by graduating class and race/ethnicity, respectively. Figures A.5 and A.6 in Appendix A present similar data with the continuation rate calculated using number of graduates who completed their ISO as the denominator (instead of all graduates). Because the ISO completion rates are high, there is little difference between the two sets of continuation rates.

Looking first at average continuation rates, 86 percent of officers in the graduating class of 2003, who were five years beyond graduation in June 2008, had just completed their ISO and were still in service as of that time. In this group, 99 percent of those who had completed their ISO were still in service in June 2008 (see Figure A.5). For the graduating class of 2002 (six years beyond graduation and one year beyond the ISO point), there was a slight drop-off in the continuation rate of graduates: Eighty-two percent remained in service one year beyond their ISO, and about three-quarters remained in service seven years beyond graduation, or two years beyond the ISO point. (Of those who had completed their ISO, the corresponding numbers were 91 percent and 81 percent; see Figure A.5.)

After that, there was a greater drop-off in terms of continuation rates, and, because ISO completion rates were high for earlier cohorts, there is little difference in the two types of continuation rates. Between 50 percent and 65 percent stayed for eight to ten years beyond graduation (three to five years beyond the ISO point), and about 40 percent remained 13–15 years beyond graduation, or eight to ten years beyond the ISO point. One point to note is that continuation is substantially affected by additional service obligations if USNA graduates become pilots or navigators.

As mentioned in Chapters Three and Four, these data suffer from an important limitation. Each data point reflects the experience of just one class. Our current data do not allow us to examine whether the intermediate outcomes of each class are similar. All we can say is that if the experiences of these graduating classes are similar over time, then we would expect about 40–50 percent of graduates to stay ten to

Figure 5.12
Percentage of Graduates Remaining in Service as of June 2008, by Gender and Graduating Class, 1993–2003, USNA

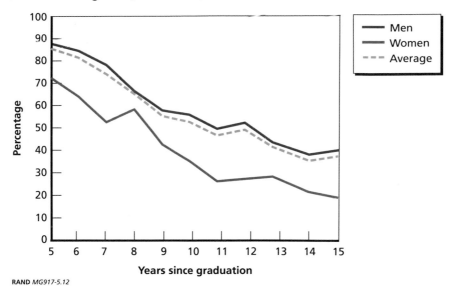

RAND *MG917-5.12*

Figure 5.13
Percentage of Graduates Remaining in Service as of June 2008, by Race/Ethnicity and Graduating Class, 1993–2003, USNA

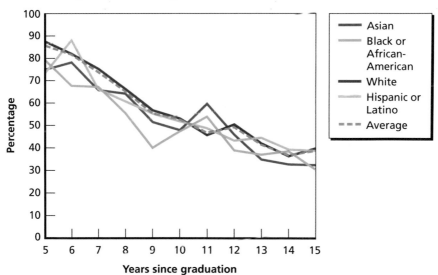

RAND *MG917-5.13*

15 years beyond graduation, or five to ten years beyond the traditional ISO point.

As shown in Figure 5.12, women had much lower continuation rates than men, and this was true of every graduating class. Immediately after completion of ISO and one year beyond ISO, we see a 16- to 20-percentage-point difference in the continuation rates of women and men (72 percent versus 88 percent and 64 percent versus 84 percent, respectively). This increased to 25 percentage points for the graduating class of 2001. However, for the class that graduated in 2000 (and so was eight years beyond graduation), 58 percent of women graduates still remained in service, compared with 66 percent of men. Generally, however, there was a 15- to 25-percentage-point difference in continuation rates. By the tenth year after ISO completion (15 years beyond graduation), only 20 percent of women graduates of the class of 1993 remained in service as of June 2008, compared with 41 percent of men.

Figure 5.13 presents continuation rates for graduating classes by race/ethnicity. The findings included the following:

- There is variability among groups in terms of continuation, with Hispanic, black, and Asian officers having higher continuation rates in particular years.
- Hispanics tended to have continuation rates similar to those of whites (with the exception of the most recent graduating class).
- Blacks in most years continued at lower rates than other groups.
- Beyond nine years (graduating class of 1995), the differences in continuation rates became considerably smaller, with only 30–40 percent of the graduates remaining in service.

Graduates Who Join the Marine Corps

USNA provided us with data on graduates who joined the Marine Corps after graduation for the graduating classes of 1996–2003. On average, about 17 percent of USNA graduates join the Marine Corps after graduation. In our sample, the total number of graduates joining the Marine Corps ranged from 154 to 165 in recent years, with the exception of the graduating class of 1999, in which the number was somewhat lower (148).

Figure 5.14 shows the percentage of USNA graduates joining the Marine Corps who were women and nonwhite. The percentage of non-whites ranged from a low of 9 percent in 1997 to a high of 24 percent in 2001, although the actual number of nonwhite graduates who joined the Marine Corps was small, ranging from 15 to 37. The percentage of women also varied over time—from a low of 8 percent in 1997 to a high of 19 percent in 2001. (Again, the actual number of women graduates joining the Marine Corps was small, ranging from 13 to 29.) For the most recent class for which we have data (2003 graduates), the percentages of graduates who joined the Marine Corps who were non-white and women were 11 percent and 13 percent, respectively.

Another way of looking at these data is to examine the propensity of women and nonwhite graduates to join the Marine Corps. Among these graduating classes, the percentage of women graduates who joined the Marine Corps ranged from 11 percent (class of 1997) to 21 percent (class of 2002), while the percentage of nonwhite graduates who joined the Marine Corps generally ranged from 16 percent (class

Figure 5.14
Percentage of USNA Graduates Who Joined the Marine Corps, by Gender and Race/Ethnicity, 1996–2003 Graduating Classes

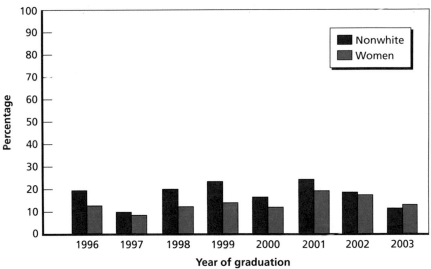

of 2000) to 21 percent (classes of 1999 and 2001). However, in the most recent graduating cohort for which we have data (class of 2003), only 10 percent of nonwhite graduates joined the Marine Corps. The data indicate that the Marine Corps attracts a representative group of USNA nonwhite and women graduates.

We cannot disaggregate outcomes for Marine Corps officers by gender or race/ethnicity because of small sample sizes. However, we can compare the average outcomes for USNA graduates who joined the Marine Corps versus the Navy. Figure 5.15 shows the ISO completion rates of the two groups of officers. Marine Corps officers had a higher rate of ISO completion, regardless of graduating class. Across all graduating cohorts, the average ISO completion rate was 97 percent for Marine Corps officers, compared with 92 percent for Navy officers. For the three most recent cohorts, the average ISO completion rate was 97 percent for Marine Corps officers and 87 percent for Navy officers. The large difference is driven by the recent decline in ISO completion rates for recent USNA graduates.

Figure 5.15
Percentage of USNA Graduates Who Joined the Navy or the Marine Corps and Completed Their ISO, 1996–2003 Graduating Classes

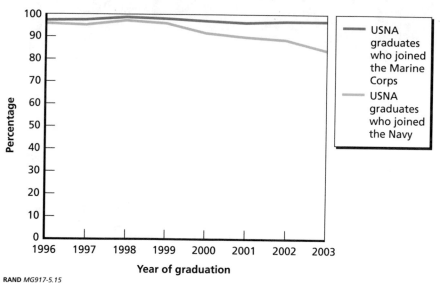

Figure 5.16 shows the percentage of Marine Corps and Navy officers who graduated from USNA between 1996 and 2003 who remained in service as of June 2008. Apart from the 2003 graduating class, Navy officers tended to have slightly higher continuation rates than Marine Corps officers in most years. Half of the Navy officers who graduated in 1996 (12 years beyond graduation, or seven years beyond the ISO decision point) were still in service in June 2008, compared with 47 percent of Marine Corps officers.

The next chapter summarizes the Army, Air Force, and Navy service action plans, provided by their respective service academies.

Figure 5.16
Percentage of USNA Graduates Who Joined the Navy or Marine Corps Remaining in Service as of June 2008, 1996–2003 Graduating Classes

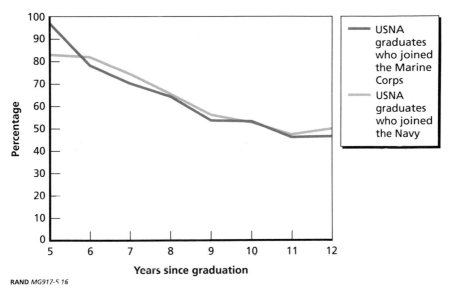

RAND *MG917-5 16*

CHAPTER SIX

Service Action Plans

Each of the military departments was asked to provide an action plan that detailed ways to improve diversity and representation. Plans were provided by each of the academies and are summarized in this chapter.

Army

The Army provided an information paper on African-American recruitment and a briefing on the same topic, which was presented to the Army Chief of Staff. USMA states that it is meeting its class composition goals for all minority demographics with the exception of African-Americans. The goal is 10–12 percent, with 6.3 percent achieved for the most recent class.

The primary minority recruitment tool is Project Outreach, which seeks to identify and nurture talented minority candidates through the admissions process, with the ultimate goal of matriculating them to West Point. Five recent West Point graduates spend a 13-month tour traveling extensively throughout selected U.S. regions to identify and nurture candidates.

Other current major programs include the following:

- Visitation program: Prospective recruits make weekend visits to USMA and the United States Military Academy Prep School.
- Metropolitan blitzes: Outreach officers and the minority-admissions officer converge on one city to mass resources.

- Representatives visit with the Congressional Black and Hispanic Caucuses to set up academy days and place cadets as interns in local and Washington, D.C., offices.
- Minority cadets participate in hometown visits and academy days.
- Cadet Calling Program: Current cadets interact with candidates via phone.

New initiatives that the USMA Directorate of Admissions is working on in an attempt to meet the goal include the following:

- increased outreach efforts with the Congressional Black Caucus members to increase nominations
- school partnerships—attempts to establish feeder schools for African-American recruits
- a minority marketing campaign through which an outside marketing firm is reviewing the entire marketing campaign with a focus on minority marketing
- an increased visitation program
- improved project outreach, such as efforts to understand school attributes that are a good predictor of cadet candidacy, to better focus efforts on the right schools
- an examination of best practices of other tier-1 institutions.

Overall, the admissions department is confident that a combination of current and new initiatives will assist in meeting recruiting goals. However, the small candidate pool of qualified African-Americans is highly sought after by other institutions and universities.

Air Force

The Assistant Secretary of the Air Force for Manpower and Reserve Affairs provided a paper and the USAFA Cadet Wing Diversity Plan from October 2007. The diversity plan articulates the superintendent's vision and direction regarding increasing diversity in the cadet wing and permanent party at USAFA. The following principles guide the action plan:

- Establish self-sustaining programs that identify diverse candidates.
- Attract highly qualified, diverse candidates to the cadet wing.
- Attract junior enlisted troops with leadership potential.
- Attract applicants with strategic language skills and/or aptitude.
- Attract first generation college/low-income/disadvantaged high school students.
- Continue to encourage or assist congressional districts with their nomination efforts.
- Continue an admissions program that gives individualized consideration to constitutionally permissible diversity factors and ensures the collection of the diversity-factor data necessary to analyze the effectiveness of diversity recruiting efforts.

The action plan is grouped into several areas that focus on the following current initiatives, among others:

- Identifying candidates: Advertise in prominent minority and urban media outlets and increase emphasis on coordinators who help identify, mentor, and evaluate diverse candidates.
- Target populations: Expand the Diversity Recruiting Division and increase cooperative efforts with other Air Force outreach programs.
- Seminars: Conduct one-week summer seminars between junior and senior years, develop grassroots information for cadets to use while at home, and expand the diversity visitation program to bring applicants to USAFA for a visit, among other efforts.
- Outreach: Conduct recruiting visits targeting high-minority-concentration and disadvantaged population centers with recent graduates; host students and parents or guests for admissions tours.
- Family, community, and other organizations: Build partnerships with national and regional organizations.

- Educational efforts: Conduct congressional workshops and staffer orientations, build and solidify relations with congressional caucuses, and organize educator orientations.
- Retention: Develop strategies to promote cadet success, provide support for cadet groups, and assist cadets of diverse backgrounds in successfully navigating the demands of USAFA.

Navy

The Superintendent of the Naval Academy stated that his number one goal for his tenure is to improve diversity. The superintendent and his leadership team are institutionalizing processes that will maintain the high quality of USNA but with a diverse brigade of midshipmen, faculty, and staff. Some progress has been made: A new diversity office, led by a senior naval officer, was created and staffed to be the single coordinating entity for all diversity efforts. The most recent classes to enter USNA and the Naval Academy Preparatory School were the most diverse classes in history, with more than 28-percent minority enrollment. The superintendent will continue to lead Naval Academy personnel in the areas of admissions outreach and midshipman retention with the goal of graduating a more diverse officer corps.

CHAPTER SEVEN

Selected Findings and Recommendations from the Literature

We conducted a brief examination of the literature on improving diversity in organizations to determine whether there were findings that would be relevant or useful to DoD and the services in developing and implementing a strategic plan to improve diversity both within and across DoD. This is by no means an exhaustive review of the literature. Nonetheless, the findings and issues raised in this chapter should be of interest.

Leadership

The one clear theme that emerged in the literature was the need for leadership and commitment to diversity at the senior levels. For example, a 2005 U.S. Government Accountability Office (GAO) report that examined expert-identified leading practices and agency examples with respect to diversity management identified the following as three of the top nine practices:

> *Top leadership commitment*—a vision of diversity demonstrated and communicated throughout an organization by top-level management.

> *Diversity as part of an organization's strategic plan*—a diversity strategy and plan that are developed and aligned with the organization's strategic plan.

Diversity linked to performance—the understanding that a more diverse and inclusive work environment can yield greater productivity and help improve individual and organizational performance. (p. 4; emphasis in original)

Similarly, the 2008 RAND monograph *Planning for Diversity* (Lim, Cho, and Curry, 2008, p. 56) stated,

[W]e recommend that the strategic planning process be top-down rather than bottom-up . . . its success depends on the leadership's ability to champion the effort, monitor its progress, and follow through on accountability measures. . . . We recommend that the Secretary personally lead an oversight committee that approves and monitors the progress of diversity initiatives. As such, we recommend that DoD form an oversight committee of top DoD leaders from a wide range of personal and professional/functional backgrounds (e.g., intelligence, combat arms, Joint Chiefs of Staff) to oversee the development of the strategic plan and its implementation, providing both insights from their vast experience and inputs from their functional communities.

Reyes (2006, p. 32), in an article for the Joint Center for Political and Economic Studies, made the same point with respect to the Army:

A final important near-term strategy is wider dissemination of the message that diversity is a critical component of a strong military force. A strategic communications plan is required to focus the U.S. Army—and those it is trying to reach—on the importance of this message. In the near term, leaders at all levels of the Army, especially at the top, must begin weaving in the message about diversity at every opportunity—in speeches, during public appearances, and at meetings and conferences. Our Army Values and Warrior Ethos are a part of almost every speech given by the Secretary of the Army, the Chief of Staff of the Army, and the Command Sergeant Major of the Army—and so they should be.

Definition of Diversity

Lim, Cho, and Curry (2008, p. x) identified three possible definitions for use by DoD:

> The first definition focuses on representation of certain groups, commonly based on U.S. Equal Employment Opportunity Commission (EEOC) categories, such as race, ethnicity, gender, and disability.

> The second definition is broader and encompasses a multitude of attributes that can influence the effectiveness of DoD in executing its mission.

> The third definition is a combination of both. It calls for prioritizing representation of certain groups and includes attributes based on DoD's needs and mission-readiness.

The authors also recommended that DoD adopt a vision based on the third definition:

> This will result in a vision that will have historical credibility and a clear "business case." Both are essential elements of an inspiring vision. Having historical credibility is important, because internal and external stakeholders—minority and female civilian employees and service members, members of Congress, and civil society at large—may perceive a vision without historical credibility as a way to avoid improving representation of minorities and women among the leadership. This perception would be reinforced by the fact that DoD's estimates indicate virtually no prospect of an increase in representation of minorities or women in the higher ranks (flag and Senior Executive Service) for the next decade, while minority populations are expected to grow significantly in the near future. . . . Having a clear business case is essential, because a vision without a clear business case will fail to instill diversity as one of the core values of DoD in the workforce. This will weaken the implementation of the strategic plan. (Lim, Cho, and Curry, 2008, pp. x–xi)

Since the publication of that report, DoD has defined *diversity* as "the different characteristics and attributes of individuals" (DoDI 1020.02, 2009), but this definition is under review. We note that the Military Leadership Diversity Commission has been tasked with coming up with a uniform definition of *diversity* (DoD, 2009). In any case, once a standardized definition has been adopted, it must be made clear and communicated to the services and the service academies.

In setting goals, a 2006 GAO report emphasized the need to ensure that the correct comparison groups are being used:

> In addition to reporting and comparing representation levels overall and in subsets of the federal workforce to the CLF [Civilian Labor Force], EEOC and OPM [Office of Personnel Management] require that agencies analyze their own workforces. However, the CLF benchmarks of representation that EEOC, OPM, and the agencies use do not differentiate between citizens and non-citizens, and therefore do not identify how citizenship affects the pool of persons qualified to work for the federal government.

Strategies

Setting a vision, mission, and goals will communicate the leadership's priorities to the rest of DoD, serving as a guide to implementation and resource allocation. Strategies to achieve these goals include

- *process strategies* that are related to operational elements, including but not limited to accessions, development, career assignments, promotion, and retention

- *enabling strategies* that involve functions that are more far-reaching in nature, such as leadership engagement, accountability, and culture.

  . . . [E]nabling strategies are necessary conditions for the success of process strategies. This is because the essence of diversity managements calls on individuals to go beyond the comfort of familiarity and uniformity. (Lim, Cho, and Curry, 2008, p. xii; emphasis in original)

Interestingly, Lim, Cho, and Curry (2008, pp. 57–58) also offered the following recommendation:

We recommend that DoD invest heavily in strategies other than those related to accessions, particularly if the chosen vision moves beyond protection of underrepresented groups. Development and retention have been overlooked thus far in many diversity efforts, which has not improved the pipeline situation for DoD. Being a closed system, DoD must retain, develop, and promote more members of diverse groups in order to achieve diversity at the very top. Moreover, it is essential that any major initiative related to leadership development must explicitly address how it will contribute to greater diversity among DoD leadership. . . . Additionally, if DoD adopts a diversity definition tied to the overall department mission that includes attributes such as language skills and cultural awareness, actions must be taken both to incentivize young recruits to attain these relevant skills and promote a career path toward the top-ranking leadership for such individuals. If DoD includes structural diversity (i.e., different components) in its definition, efforts must be made to effectively integrate all components (active duty, reserve, and civilian) encouraging a better understanding of each component's unique contribution to the mission.

GAO (2005, p. 4) pointed to the importance of succession planning and defined it as an "ongoing, strategic process for identifying and developing a diverse pool of talent for an organization's potential future leaders."

Reyes (2006, p. 31) points to some of the steps taken by the Navy as important in increasing the number of black men recruited and accessed into the Army:

We may also look to the U.S. Navy for examples of ways to improve black accession rates. In recruitment and accessions, the Navy has increased its budget for implementing strategies designed to increase diversity. The Navy has established a Navy Office of Community Outreach (NAVCO), which is already coordinating activities with the NAACP [National Association for the Advancement of Colored People] and HENAAC (the Hispanic Engineers National Achievement Awards Corporation). A Diversity "Recruiters of the Year" award has been established. The Navy has increased enlisted diversity recruit quality five years in a row and quarterly updates on the diversity of officer recruitment are provided to the leadership. In addition, the U.S. Naval Academy (USNA) has established a Minority Outreach Coordinator.

He also emphasizes the need to broaden black officers' experience by increasing their representation in high-profile career-enhancing jobs:

> An important action for the Army to take is to increase black officer representation in the high-profile career-enhancing jobs (e.g., line unit commander, operations officer, executive officer, aide-de-camps, etc). Leaders must ensure that opportunities to work in these career-enhancing positions exist for black officers as these jobs become available. This leadership responsibility is important because it addresses a situation in which many black officers find themselves. (Reyes, 2006, p. 32)

One way of accomplishing this is through mentorship as well as focusing on retention:

> Mentoring plays a vital role in the effort to help more black officers attain career-enhancing jobs. The guidance of a mentor is important to ensuring that officers stay focused on pursuing the critical and challenging jobs. While doing well on any assignment is important, a strong performance in a job that is understood to be challenging and critical sets an officer apart and distinguishes him/her as potential material for service in the senior ranks. In many cases, if an officer fails to receive this kind of mentoring early in his/her career, that officer strives for a job where success is more easily attained, rather than a demanding and career-enhancing job that could lead to greater responsibility in the future and clear opportunities for promotion.

. . . [T]he Army is making an effort to provide mentorship in some capacity. However, it is my belief that a formal mentorship program is required. Senior officers should be required to have protégées and young officers should be expected to seek out senior officers as mentors. The optimum scenario would be for each officer to have multiple mentors from different cultural backgrounds over the course of his/her career. This would allow officers to gain a wealth of knowledge from the diverse experiences and cultural perspectives of their mentors. (Reyes, 2006, p. 32)

A final necessary long-term strategy concerns retention rates. The Army must focus its retention efforts so that it establishes a talent pool of quality black officers and a healthy pipeline to the senior ranks. At the same time, the Army must prove itself capable of competing with a civilian sector that has successfully recruited workers from the Army's ranks. To be competitive, the Army must enhance its attractiveness in areas such as salary and quality of life. In addition, the Army must communicate the advantages of a career in the Army, including the value of military experience. (Reyes, 2006, p. 34)

Reyes (2006, p. 33) also highlighted the importance of engaging the community:

One critical action that the Army should take is to engage the community as part of a long-term strategy to boost the number of blacks recruited by commissioning sources. Family members, religious leaders, and local politicians can all play an influential role in supporting recruitment efforts. First, however, the Army must reach out to these communities. Our ROTC units are doing an outstanding job, but they cannot do it alone.

GAO (2006, pp. 24–25) described some interesting strategies used by the National Aeronautics and Space Administration (NASA) to recruit Hispanics:

Part of NASA's strategy to recruit Hispanics centers on increasing educational attainment, beginning in kindergarten and continuing into college and graduate school, with the goal of attracting students into the NASA workforce and aerospace community. NASA centers sponsor, and its employees participate in, mentoring, tutoring, and other programs to encourage Hispanic and other students to pursue careers in science, engineering, technology, and math. For example, the Marshall Space Center in Huntsville, Alabama, annually sponsors a Hispanic Youth Conference attended by students from across Alabama that includes workshops on leadership development and pursuing NASA career fields and provides opportunities to establish mentoring relationships. NASA also provides grants to fund educational support programs including in locations where there are high concentrations of Hispanics. For example, the Ames Research Center in Moffett Field, California, provided a grant for the development and implementation of a K–12 technology-awareness program designed to expose students to NASA and higher education through competitive team activities based on key aeronautic concepts. The program has been implemented in schools throughout California that have a high percentage of Hispanic students. Various centers also participate in high school and college internship programs, such as the Summer High School Apprenticeship Research Program where high school students spend 8 weeks working with engineers on scientific, engineering, mathematical, and technical projects. NASA centers also provide scholarships and research grants. For example, Ames provides scholarships to Hispanic college students at a community college and the Dryden Flight Research Center sponsors fellowships for students in engineering and science to continue their graduate studies. In addition, NASA has recently developed the Motivating Undergraduates in Science and Technology scholarship program designed to stimulate a continued interest in science, technology, engineering, and mathematics.

Aronson's recommendations echo several of the ones already mentioned:

Define carefully and accurately the job selection criteria, such as the particular skills and abilities required, before the selection process begins.

Partner with minority associations and educational institutions, participate in minority career festivals, and advertise in minority media.

Develop educational outreach programs, such as scholarships, internships, and work/study programs. Explore community involvement options that bring the company goodwill and that open lines of communication.

Work on eliminating barriers to hiring minorities, and communicate to all stakeholders the company's ongoing efforts to expand the candidate. (Aronson, 2002, p. 60)

Evaluation and Metrics to Guide Progress

The literature emphasizes the need for careful monitoring of progress towards the achievement of goals. Nelson, Cho, and Curry (2008, pp. xiii–xiv) noted,

Evaluation serves as the link between strategic planning and implementation by tracking the progress of on-the-ground efforts and informing accountability processes. Metrics for evaluation ought to be derived from the vision, but this is not currently the case with diversity because the field lacks appropriate metrics. Various metrics are available or under development to measure

- diversity in a group
- organizational climate
- intermediate (process) and final outcomes.

Many of these metrics are untested or not feasible to apply in the field. Most organizations, including DoD and its components (the Military Departments and the Fourth Estate), default to measurement of demographic representation and climate surveys, even though they have adopted a broad vision of diversity that goes beyond demographic diversity. This mismatch between the vision and metrics results in confusion and dilutes the impact of diversity initiatives. A more strategic approach for DoD would involve (1) determining what needs to be measured according to the leadership's vision and mission for diversity and (2) employing and/or developing metrics that support the vision and mission. Head counting, for example, is appropriate for measuring representations of certain groups, but it will not completely capture the most important aspects of a diversity vision that emphasizes inclusion. DoD must be creative and innovative when developing new metrics that focus on mission-readiness.

A 2007 GAO report was critical of the data provided by the DMDC and noted that efforts should be made to ensure accuracy in the data. Without accurate information, it is difficult to measure whether agency objectives are being met:

> We found the information that DMDC provided to us on the number of officers accessed from DOD's various commissioning programs to be insufficiently reliable for use in our January 2007 report. Government auditing standards, which are applicable to all federal agencies including DOD, require that data be valid and reliable when the data are significant to the auditor's findings. More specifically, federal internal control standards require that data control activities, such as edit checks, verification, and reconciliation, be conducted and documented to help provide reasonable assurance that agency objectives are being met. We found discrepancies when we compared the DMDC-provided information on the number of officers accessed from DOD's commissioning programs (the academies, ROTC, and [Officer Candidate Schools/Officer Training Schools]) to information provided by the services. (GAO, 2007, p. 2)

The 2005 GAO report also listed measurement and accountability as two of the top nine practices:

> *Measurement*—a set of quantitative and qualitative measures of the impact of various aspects of an overall diversity program.

> *Accountability*—the means to ensure that leaders are responsible for diversity by linking their performance assessment and com-

pensation to the progress of diversity initiatives. (GAO, 2005, p. 4; emphasis in original)

Summary

Our review of the literature pointed to several steps that DoD could adopt to support the services in their efforts to improve diversity, both in the academies and in the officer corps. First, it is important to communicate DoD's definition of *diversity* and to use it to set goals for diversity that are aligned with DoD's overall mission and to measure progress against those goals. Second, DoD should conduct an assessment of the current metrics to determine whether they are appropriate and should develop new metrics, if necessary. Third, DoD should make clear to the services that diversity management is a priority for the entire organization and has the backing of the highest level of DoD leadership—not merely the personnel community. One way of emphasizing this would be to link performance assessment of leaders to goal achievement, as discussed in GAO (2005). Fourth, it is important to focus efforts not simply on accessing a more diverse group of officers but on increasing career retention among them. Without improving career retention rates, it is not possible to improve diversity at the highest levels of leadership. Improving career retention may require an understanding of the choice of occupations and assignments that tend to affect promotion opportunities and, ultimately, retention and increased efforts to provide mentorship and counseling to junior officers.

CHAPTER EIGHT

Conclusions and Recommendations

We were tasked with the following: (1) examining the demographic profile of service academy entrants and changes over time, (2) analyzing various short- and longer-term outcomes for academy entrants and graduates to see whether and how these varied by gender and race/ethnicity, (3) summarizing the action plans of the academies to improve diversity, and (4) providing recommendations to DoD and the services to support and further these efforts. The study was a short-term effort and limited in scope to descriptive analyses of materials provided by the academies.

Findings

Demographic Profile of Entrants

Comparing the three earliest entry cohorts (1992–1994) with the three most recent cohorts for which we have data (2007–2009), we found that the percentage of women increased from 12 percent to 16 percent in USMA and from 15 percent to 21 percent in USAFA and USNA.

The percentage of nonwhites also increased over time in the three academies:

- In USMA, the percentage of nonwhites increased from 16 percent in the 1992–1994 entering classes to 23 percent in the 2007–2009 entering classes. Of the 2007–2009 entering classes, 6 percent were black or African-American, 9 percent were Hispanic or Latino, 7 percent were Asian, and 1 percent were American Indian or Alaska Native.

- In USAFA, the percentage of nonwhites increased from 18 percent to 22 percent over the same period. In the most recent classes, 6 percent of academy entrants were black or African-American, 8 percent were Hispanic or Latino, 7 percent were Asian, and 1 percent were American Indian or Alaska Native.
- USNA experienced a similar increase in the percentage of nonwhites (from 18 percent to 22 percent). However, the percentage of nonwhites in the 2009 entering class rose to 28 percent. In the most recent classes, 5 percent of academy entrants were black or African-American, 12 percent were Hispanic or Latino, 4 percent were Asian, and less than 1 percent were American Indian or Alaska Native.

First-Year Completion and Graduation

USMA. The percentage successfully completing the first year among the USMA entry cohorts increased from 84 percent in the earliest classes (1992–1994) to 91 percent in the most recent classes (2006–2008). Women, in particular, increased their first-year completion rate by 10 percentage points. Every racial/ethnic group increased their first-year completion rate by 3–10 percentage points.

USMA graduation rates remained relatively constant over time (76–77 percent). Again, women increased their graduation rates by 5 percentage points (from 69 percent in the 1992–1994 entry cohorts to 74 percent in the 2003–2005 entry cohorts). With one exception (Asians), all racial/ethnic groups either maintained or slightly increased their graduation rates over time. The graduation rate conditional on successfully completing the first year declined from 92 percent in the earliest cohorts (1992–1994) to 87 percent in the 2003–2005 entering cohorts.

USAFA. The first-year completion rate increased from 81 percent in the three earliest cohorts to 85 percent in the three most recent cohorts. Women essentially maintained their first-year completion rate. All racial/ethnic groups increased their first-year completion rate—

Hispanics by 2 percentage points, blacks by 8 percentage points, and Asians by 6 percentage points.

The overall graduation rate also increased over time, from 72 percent in the three earliest cohorts to 76 percent in the most recent cohorts. Women posted a gain of 3 percentage points, while blacks and Hispanics increased their graduation rate by 2 percentage points. Asians and whites experienced larger increases of 6–7 percentage points.

As was true for USMA, the conditional graduation rate declined slightly by 2 percentage points between the two periods; there were also declines in conditional graduation rates among women, blacks, and Hispanics.

USNA. The first-year completion rate increased by 6 percentage points (from 88 percent in the three earliest cohorts to 94 percent in the three most recent cohorts). Women increased their first-year completion rate by 8 percentage points to 91 percent in the most recent cohorts. All racial/ethnic groups increased their first-year completion rate, and there was little difference in completion rates across the various groups.

The graduation rate also increased over time, from 78 percent in the three earliest cohorts to 85 percent in the most recent cohorts—the highest among the three academies. Women increased their graduation rate substantially, from 70 percent to 85 percent over this period, bringing it on a par with men. Hispanics also experienced a marked increase in graduation rates—from 68 percent to 81 percent—while blacks increased their graduation rates by 8 percentage points to 75 percent. Asians and whites also posted increases of 7–8 percentage points in graduation rates.

Women in earlier cohorts had higher conditional graduation rates than men; in the most recent cohorts, the rate for men increased slightly by 2 percentage points to 91 percent, bringing them on a par with women. The conditional graduation rates of blacks and Hispanics improved by 8–10 percentage points to 85 percent and 89 percent, respectively.

ISO Completion

USMA. The rate of graduates' ISO completion increased by 9 percentage points across time (from 82 percent for the 1993–1995 graduating classes to 91 percent for the 2001–2003 graduating classes). About 91 percent of women graduates completed their ISO in the most recent cohorts, compared with 79 percent in the earliest cohorts. The increase in ISO completion over the same period ranged from 8 percentage points for blacks to 12 percentage points for Asians.

USAFA. Among graduates, the ISO completion rate declined from 90 percent for the earliest cohorts to 82 percent for the more recent cohorts. The ISO completion rate declined markedly for women (from 86 percent to 69 percent) and for blacks (from 90 percent to 76 percent) and Asians (from 93 percent to 80 percent).

USNA. Among graduates, the ISO completion rate declined from 95 percent for the earliest cohorts to 89 percent for the more recent cohorts. As with USAFA, the ISO completion rate declined markedly for women (from 89 percent to 75 percent). There were also large declines of 8–13 percentage points in ISO completion rates among all nonwhite groups.

USNA graduates who joined the Marine Corps had a higher rate of ISO completion than those who joined the Navy, regardless of graduating class. In the three most recent cohorts (2001–2003), the average rate of ISO completion was 97 percent for Marine Corps officers, compared with 87 percent for Navy officers.

Exogenous factors—such as reductions in force or competition from the civilian economy—are likely to affect retention and ISO completion, and this may help explain the decline. An analysis of the effects of such factors was beyond the scope of this study.

Continuation Rates

USMA. Of the graduating class of 2003, 73 percent remained in service as of June 2008, when they had just completed their ISO.[1] For the graduating class of 2002, there was a sharp drop-off in continuation

---

[1]   As discussed earlier, our data on continuation rates are limited because each data point represents the experience of only one graduating class.

rates: Only 52 percent remained in service one year beyond their ISO. If the experiences of the graduating classes are similar over time, then we would expect one-third of graduates to remain in the military for seven to ten years beyond their ISO (12–15 years beyond graduation).

Women had much lower continuation rates than men, and this was true of every graduating class. The continuation rates of nonwhite graduates tended to be more variable, but, in general, they—especially Hispanics—appeared to continue at higher rates than whites six to nine years beyond their ISO.

USAFA. Of the graduating class of 2003, 82 percent remained in service as of June 2008, when they had just completed their ISO. We found that, over time, about half of the graduates stayed six to ten years beyond their ISO. This continuation rate is higher than that of USMA, largely because some graduates incur additional service obligations as pilots.

Women had much lower continuation rates than men, and this was true of every graduating class. Blacks in most years continued in service at lower rates than other groups.

USNA. Of the graduating class of 2003, 86 percent remained in service as of June 2008, when they had just completed their ISO. Of the graduating class of 2002, 82 percent remained in service one year beyond their ISO. Over time, 40–50 percent of graduates stayed seven to ten years beyond their ISO.

Women had lower continuation rates than men, and this was true of every graduating class. Hispanics generally continued in service at rates similar to those of whites, and blacks in most years continued at lower rates than other groups.

Statistically Significant Differences in Selected Outcomes Across All Cohorts

Table 8.1 summarizes statistically significant differences among groups for the selected outcomes across the three service academies. As noted earlier, for the sake of brevity, we refer to statistically significant differences as *significant differences*. These data are across all years; they do not simply compare the earliest and most recent cohorts, so some of these differences may not hold in future cohorts if outcomes for selected groups continue to improve and differences between groups become smaller.

Table 8.1
Statistically Significant Estimated Differences in Selected Outcomes Across All Years, by Gender and Race/Ethnicity

Characteristic	First-Year Completion	Graduation from the Academy	Completion of ISO
Women, compared to men:			
USMA	Lower	Lower	—
USAFA	*Lower*	—	Lower
USNA	Lower	Lower	Lower
Compared to whites:			
USMA			
Asian	Higher	Higher	Higher
Black or African-American	—	Lower	*Lower*
Hispanic or Latino	—	Lower	—
USAFA			
Asian	Higher	—	—
Black or African-American	—	—	Lower
Hispanic or Latino	—	Lower	—
USNA			
Asian	—	—	*Lower*
Black or African-American	—	Lower	Lower
Hispanic or Latino	*Lower*	Lower	*Lower*

NOTE: All cell entries represent statistically significant differences relative to the reference group. Entries in italics represent those where the odds ratio was modest or the confidence interval was wide (the upper or lower limit was close to 1), indicating a weak effect.

— indicates no significant difference between groups.

- In seven of the nine outcomes considered here, across all cohorts, women's outcomes were significantly different and lower than those of men. However, in all three academies, as noted earlier, women improved their first-year completion and graduation rates to the point at which the differences between men and women were substantially reduced or eliminated altogether. However, in recent cohorts, USAFA and USNA women graduates had markedly lower ISO completion rates than men.
- Asians generally had outcomes similar to or better than those of whites. However, in recent cohorts, Asians graduating from USAFA and USNA had lower rates of ISO completion than whites.
- Blacks had significantly lower graduation rates than whites in USMA and USNA, and, despite recent increases, this continues to be the case. In recent cohorts, the differences in graduation rates were 3 percentage points in USMA, 5 percentage points in USAFA, and 11 percentage points in USNA. Blacks also tended to have lower ISO completion rates, and the differences were even more marked in recent cohorts.
- Hispanics had significantly lower graduation rates than whites across all three academies. However, in recent cohorts, they closed the gap to 1 percentage point in USMA, 2 percentage points in USAFA, and 5 percentage points in USNA.

Thus, while looking across cohorts is useful, it is important to consider the experience of the more recent cohorts to determine priorities for investing resources to improve outcomes for diverse groups.

Comparing Academy Graduation Rates to Graduation Rates of "Very Selective" Four-Year Institutions

We compared the graduation rates of the service academies to those of "very selective" civilian four-year institutions, using data published by NCES. Two points to note regarding the comparisons: First, the civilian institution data shown are for the freshman entering class of 1998, while the service academy data are aggregated across the 1992–2005 entering classes; second, entrants typically graduate from the service academies in four years, so we are comparing four-year graduation rates to six-year graduation rates for civilian institutions. In the 1998 freshman class, the percentage of women enrolled in very selective civilian four-year institutions (n = 117) was 57 percent, much higher than the 16–21 percent enrolled in the most recent academy entry cohorts. The percentage of nonwhites enrolled in the freshman class, however, was similar between the civilian and military institutions—23 percent in the civilian institutions compared with 22–23 percent in the most recent academy entry cohorts.[2]

Figure 8.1 shows the graduation rates by race/ethnicity and gender for very selective four-year institutions and for the service academies. Academy graduation rates are higher than those in comparable civilian four-year institutions, on average and across all racial/ethnic groups. For example, 72 percent of blacks graduated from the service academies, on average, compared with 60 percent who attended four-year civilian institutions—a substantial 12-percentage-point gap in graduation rates. We also noted earlier that graduation rates for the most recent cohorts entering the academy (2003–2005) have increased, so, if this improvement is sustained, the gap in graduation rates may be even larger.

There is a 3-percentage-point difference in the graduation rates of women (74 percent versus 77 percent in the academies and civilian institutions, respectively) and a 5-percentage-point difference in the graduation rates of men (78 percent versus 73 percent). However, while women have lower graduation rates than men in the academies, the opposite is true in the civilian institutions. We should note that the difference in graduation rates between men and women in the most recent graduating cohort declined to 1 percentage point.

---

[2] Using data only for 1998 academy entry cohort, we find that the percentage of women was slightly smaller (15–16 percent), as was the percentage of nonwhites (18–19 percent), than in the most recent cohorts.

Figure 8.1
Six-Year Graduation Rates in Very Selective Four-Year Institutions, 2004,
and the Service Academies, 2003–2005, by Race/Ethnicity and Gender

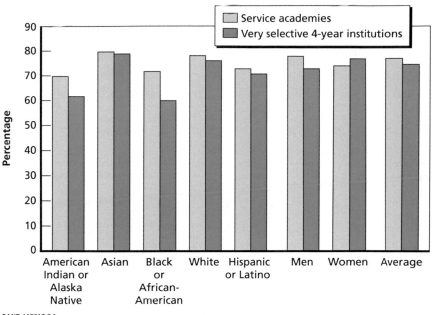

RAND *MG917-8.1*

Service Action Plans

Each of the military departments has action plans detailing ways to improve diversity and representation.

USMA. The primary minority recruitment tool is Project Outreach, which seeks to identify and nurture talented minority candidates through the admissions process with the ultimate goal of matriculating them to West Point. Other programs include weekend visits of prospective recruits to USMA and the United States Military Academy Prep School; visits with the Congressional Black and Hispanic Caucuses to set up academy days and place cadets as interns in local and Washington, D.C., offices; minority cadets' participation in hometown visits and academy days; and the Cadet Calling Program, through which current cadets interact with candidates via phone. Several other new initiatives are under way, including examining best practices at other tier-1 institutions.

USAFA. To identify candidates, USAFA will advertise in prominent minority and urban media outlets and increase emphasis on coordinators who help identify, mentor, and evaluate diverse candidates. Among other initiatives, it proposes to offer one-week summer seminars between junior and senior years, expand the diversity visitation program to bring applicants to USAFA for a visit, and provide support to cadets of diverse backgrounds to help ensure their success.

USNA. A new diversity office, led by a senior naval officer, was created and staffed to be the single coordinating entity for all diversity efforts.

Recommendations for DoD to Support Service Efforts to Improve Diversity

The action plans adopted by the service academies encompass several specific strategies. At a higher level, DoD and the services need to take steps both to support these plans and to ensure that the plans are linked to the larger DoD vision and goals. Specifically, the Office of the Secretary of Defense should pursue the following initatives:

- Review and communicate DoD's definition of *diversity.*
- Determine what needs to be measured according to the leadership's vision and mission for diversity and the best metrics for that purpose.
- Review goals for diversity and ensure that they are aligned with DoD's overall mission.
- Emphasize that diversity management is a priority for the entire organization and has the backing of the highest level of DoD leadership, not merely the personnel community.
- Focus efforts not simply on accessing a more diverse group of officers but on increasing career retention of these officers.

Continuation Rates of Graduates Who Completed Their ISO, 1993–2003 Graduating Classes

This appendix displays additional charts for the percentage of 1993–2003 graduates who continued in service as of June 30, 2008. As noted earlier, we calculated continuation rates in two ways: Chapters Three through Five showed the continuation rates of graduates; here, we show continuation rates of graduates who completed their ISO. Thus, the x-axis shows years beyond ISO completion, ranging from zero years (for the 2003 graduating class) to ten years (for the 1993 graduating class). Table 3.4 in Chapter Three provided a crosswalk between the year of graduation and years beyond ISO completion. Note that the x-axis is in reverse order in terms of graduating classes—the most recent cohorts are to the left and the earliest cohorts are to the right on the axis.

Figures A.1 and A.2 present continuation rates of USMA graduates who completed their ISO for zero to ten years beyond the ISO completion point by gender and race/ethnicity, respectively. Figures A.3 and A.4 show similar data for USAFA graduates, and Figures A.5 and A.6 present data on USNA graduates.

Figure A.1
Percentage of Graduates Who Completed Their ISO and Remained in Service as of June 2008, by Gender and Years Beyond ISO Completion, 1993–2003 Graduating Classes, USMA

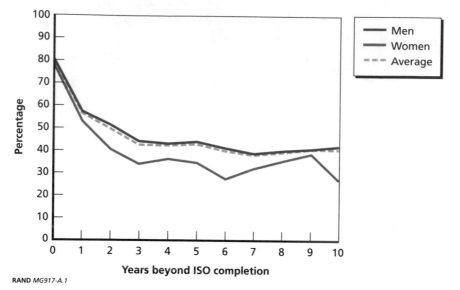

**Figure A.2
Percentage of Graduates Who Completed Their ISO and Remained
in Service as of June 2008, by Race/Ethnicity and Years Beyond ISO
Completion, 1993–2003 Graduating Classes, USMA**

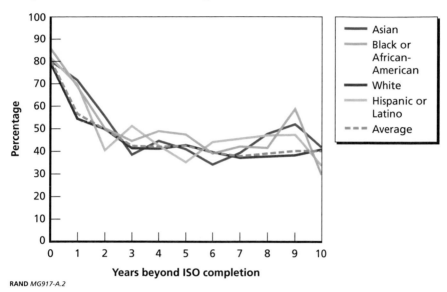

RAND *MG917-A.2*

Figure A.3
Percentage of Graduates Who Completed Their ISO and Remained in Service as of June 2008, by Gender and Years Beyond ISO Completion, 1993–2003 Graduating Classes, USAFA

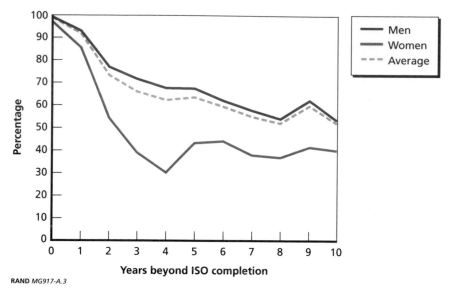

RAND *MG917-A.3*

Figure A.4
Percentage of Graduates Who Completed Their ISO and Remained
in Service as of June 2008, by Race/Ethnicity and Years Beyond ISO
Completion, 1993–2003 Graduating Classes, USAFA

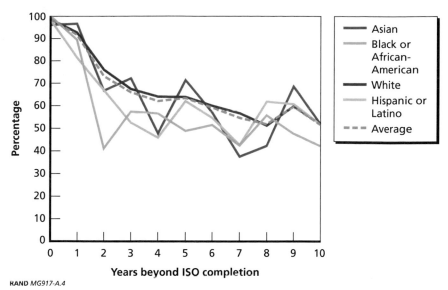

RAND *MG917-A.4*

Figure A.5
Percentage of Graduates Who Completed Their ISO and Remained in Service as of June 2008, by Gender and Years Beyond ISO Completion, 1993–2003 Graduating Classes, USNA

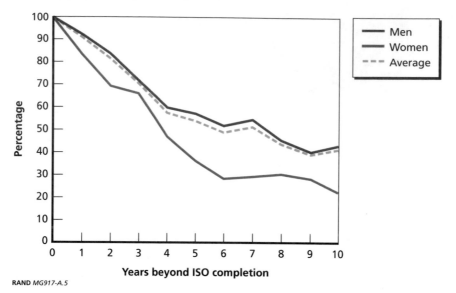

Figure A.6
**Percentage of Graduates Who Completed Their ISO and Remained
in Service as of June 2008, by Race/Ethnicity and Years Beyond ISO
Completion, 1993–2003 Graduating Classes, USNA**

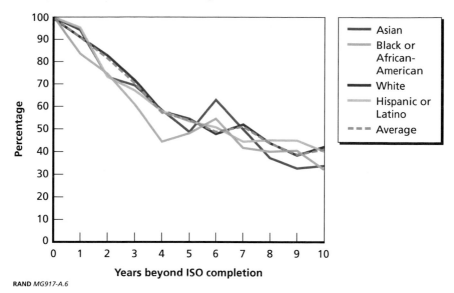

Active-Duty Service Obligations

Initial active-duty service obligations for the several commissioning sources are mandated by law and policy. Moreover, other laws or policies require service obligations for additional training opportunities. Table B.1 shows the obligations for selected educational and training opportunities.

The data in this monograph that reflect completion of initial active-duty service obligation measure only the obligation based on academy education. Moreover, the data on percentages remaining in service are likely biased upward for the Air Force and the Navy to the extent that large proportions of their graduates go on to pilot or other additional training and incur additional service obligations. Therefore, we caution that comparisons across the services on this measure would not be valid.

Table B.1
Selected Active-Duty Service Obligations

Commissioning Source	Obligation
Army ROTC	4 years (3 years if nonscholarship)
Navy ROTC	4 years
Air Force ROTC	4 years
USMA	5 years
USNA	5 years
USAFA	5 years
Additional Training	**Obligation**
Army aviator	6 years from completion of training
Navy, nuclear	5 years from completion of training
Navy pilot	8 years from completion of training
Navy naval flight officer	6 years from completion of training
Air Force pilot	10 years from completion of training
Air Force navigator	6 years from completion of training

NOTE: Additional obligations for other types of training or education (e.g., health professions) are not included in the table.

References

Aronson, David, "Managing the Diversity Revolution: Best Practices for 21st Century Business," *Civil Rights Journal*, Vol. 6, No. 1, Winter 2002, pp. 46–63.

Cunningham, Alisa F., *Changes in Patterns of Prices and Financial Aid*, Washington, D.C.: National Center for Education Statistics, 2005. As of October 9, 2009:
http://nces.ed.gov/pubSearch/pubsinfo.asp?pubid=2006153

Deputy Under Secretary of Defense for Military Personnel Policy, "Military Personnel Policy," Web page, undated. As of October 9, 2009:
http://www.defenselink.mil/prhome/mpp.html

DoD—*see* U.S. Department of Defense.

DoDI—*see* U.S. Department of Defense Directive.

GAO—*see* U.S. Government Accountability Office.

Horn, Laura, *Placing College Graduation Rates in Context: How 4-Year College Graduation Rates Vary with Selectivity and the Size of Low-Income Enrollment*, Washington, D.C.: National Center for Education Statistics, 2006. As of October 9, 2009:
http://nces.ed.gov/Pubsearch/pubsinfo.asp?pubid=2007161

Hosek, Susan D., Peter Tiemeyer, M. Rebecca Kilburn, Debra A. Strong, Selika Ducksworth, and Reginald Ray, *Minority and Gender Differences in Officer Career Progression*, Santa Monica, Calif.: RAND Corporation, MR-1184-OSD, 2001. As of October 9, 2009:
http://www.rand.org/pubs/monograph_reports/MR1184/

Lim, Nelson, Michelle Cho, and Kimberly Curry, *Planning for Diversity: Options and Recommendations for DoD Leaders*, Santa Monica, Calif.: RAND Corporation, MG-743-OSD, 2008. As of October 12, 2009:
http://www.rand.org/pubs/monographs/MG743/

National Center for Education Statistics, "About IPEDS," Web page, undated(a). As of October 9, 2009:
http://nces.ed.gov/ipeds/about/

—————, "IPEDS Survey Components and Data Collection and Dissemination Cycle," Web page, undated(b). As of October 9, 2009:
http://nces.ed.gov/ipeds/resource/survey_components.asp

NCES—*see* National Center for Education Statistics.

Office of Management and Budget, "Revisions to the Standards for the Classification of Federal Data on Race," October 30, 1997. As of October 9, 2009:
http://www.whitehouse.gov/omb/rewrite/fedreg/ombdir15.html

OMB—*see* Office of Management and Budget.

Reyes, Anthony D., *Strategic Options for Managing Diversity in the U.S. Army*, Washington, D.C.: Joint Center for Political and Economic Studies, 2006.

U.S. Department of Defense, notice of establishment of U.S. Department of Defense federal advisory committee, *Federal Register*, Vol. 74, No. 15, Doc. E9-1423, January 26, 2009.

U.S. Department of Defense Directive 1020.02, Diversity Management and Equal Opportunity (EO) in the Department of Defense, February 5, 2009.

U.S. Government Accountability Office, *Diversity Management: Expert-Identified Leading Practices and Agency Examples*, Washington, D.C., GAO-05-90, January 2005.

—————, *The Federal Workforce: Additional Insights Could Enhance Agency Efforts Related to Hispanic Representation*, Washington, D.C., GAO-06-832, August 2006.

—————, *Military Personnel: DMDC Data on Officers' Commissioning Programs Is Insufficiently Reliable and Needs to Be Corrected*, Washington, D.C., GAO-07-372R, March 2007.

United States Code, Title 10, Section 4342, Cadets: Appointment, Numbers, Territorial Distribution [United States Military Academy].

United States Code, Title 10, Section 6954, Midshipmen: Number [United States Naval Academy].

United States Code, Title 10, Section 9342, Cadets: Appointment, Numbers, Territorial Distribution [United States Air Force Academy].

United States Air Force Academy, "Eligibility," Web page, undated. As of October 9, 2009:
http://www.academyadmissions.com/#Page/Admissions_Officers_Eligibility

United States Military Academy, "FAQs About Admission," Web page, undated. As of October 9, 2009:
http://admissions.usma.edu/FAQs/faqs_admission.cfm

United States Naval Academy, "Basic Requirements for Eligibility," Web page, undated. As of October 9, 2009:
http://www.usna.edu/admissions/steps2.htm

USAFA—*see* United States Air Force Academy.

USMA—*see* United States Military Academy.

USNA—*see* United States Naval Academy.